D1606425

PHYSICAL RELATIVITY

Physical Relativity

Space-time Structure from a Dynamical Perspective

HARVEY R. BROWN

CLARENDON PRESS · OXFORD

OXFORD
UNIVERSITY PRESS

Great Clarendon Street, Oxford OX2 6DP

Oxford University Press is a department of the University of Oxford.
It furthers the University's objective of excellence in research, scholarship,
and education by publishing worldwide in

Oxford New York

Auckland Cape Town Dar es Salaam Hong Kong Karachi
Kuala Lumpur Madrid Melbourne Mexico City Nairobi
New Delhi Shanghai Taipei Toronto

With offices in

Argentina Austria Brazil Chile Czech Republic France Greece
Guatemala Hungary Italy Japan Poland Portugal Singapore
South Korea Switzerland Thailand Turkey Ukraine Vietnam

Published in the United States
by Oxford University Press Inc., New York

First published 2005

British Library Cataloguing in Publication Data
Data available

Library of Congress Cataloging in Publication Data
Brown, Harvey R.
Physical relativity : space-time structure from a dynamical perspective / Harvey R. Brown.
p. cm.
Includes bibliographical references and index.
1. Special relativity (Physics) 2. Kinematic relativity. 3. Space and time.
4. Einstein, Albert, 1879–1955. 5. Lorentz, H. A. (Hendrik Antoon), 1853–1928. I. Title.
QC173.65.B76 2005 530.11—dc22 2005023506

Typeset by Newgen Imaging Systems (P) Ltd., Chennai, India
Printed in Great Britain
on acid-free paper by
Biddles Ltd., King's Lynn, Norfolk

ISBN 0–19–927583–1 978–0–19–927583–0

1 3 5 7 9 10 8 6 4 2

To Maita, Frances, and Lucas

When we have unified enough certain knowledge, we will understand who we are and why we are here.

If those committed to the quest fail, they will be forgiven. When lost, they will find another way. The moral imperative of humanism is the endeavour alone, whether successful or not, provided the effort is honorable and failure memorable.

Edward O. Wilson, *Consilience*

Believe those who seek the truth.
Doubt those who find it.

Saying on refrigerator magnet

Preface

As I write, the centennial of Einstein's *annus mirabilis*, and in particular of his great 1905 paper on the electrodynamics of moving bodies, is upon us. Einstein thought that the principles of the special theory of relativity would be as robust and secure as those of thermodynamics, and both the special and general theories have undoubtedly borne the test of time. Each theory in its own right is a triumph.

However confident Einstein was in the solidity of special relativity, there was nonetheless a vein of doubt running through his writings—culminating in his 1949 *Autobiographical Notes*—concerning the way he formulated the theory in 1905. It is clear, to me at least, that Einstein was fully conscious right from the beginning that there were two routes to relativistic kinematics, and that as time went on the appropriateness of the route he had chosen, which he felt he *had* to choose in 1905, was increasingly open to question. In his acclaimed 1982 scientific biography of Einstein, Abraham Pais noted with disapproval that as late as 1915, H. A. Lorentz, the contemporary physicist Einstein revered above all others, was still concerned with the dynamical underpinnings of length contraction. 'Lorentz never fully made the transition from the old dynamics to the new kinematics.' There is a sense, and an important one, in which neither did Einstein.

A small number of other commentators have, over the intervening years, voiced similar misgivings about the standard construal of the theory, whether in the 1905 formulation or in its geometrical rendition by Minkowski and others following him. It seems to me that the alternative, so-called 'constructive' route to space-time structure deserves more discussion, and in particular its significance in general relativity needs to be examined in more detail.

In fact, there are essentially two competing versions of the constructive account, and in this book I will defend what might be called the 'dynamical' version which contains an echo of some key aspects of the thinking of Hendrik Lorentz, Joseph Larmor, Henri Poincaré, and particularly George F. FitzGerald prior to the sudden explosion on the scene of Einstein. (I feel, from dire experience, I must emphasize from the outset that this approach does *not* involve postulating the existence of a hidden preferred inertial frame! The approach is *not* a version of what is sometimes called in the literature the *neo-Lorentzian interpretation* of special relativity.) The main idea appears briefly in the writings of Wolfgang Pauli and Arthur Eddington, and in a more sustained fashion in the work of W. F. G. Swann, L. Jánossy, and J. S. Bell. I have been promoting it in papers over the last decade or so, some of which were the result of a stimulating collaboration with Oliver Pooley. In a nutshell, the idea is to deny that the distinction Einstein made in his 1905 paper between the kinematical and dynamical parts of the discussion is a fundamental one, and to assert that relativistic phenomena like length contraction and time dilation are in the last analysis the result of structural properties of the quantum theory of matter. Under this construal, special relativity does not amount to a *fully* constructive theory, but nor is it a fully fledged principle theory based on phenomenological principles. Now according to a competing, fully constructive view, and one dominant within at least the philosophical literature over the last three decades or so, the basic explanation of these kinematical effects is that rods and clocks are embedded in Minkowski space-time, with its flat pseudo-Euclidean metric of Lorentzian signature. This geometric structure, purportedly left behind even if *per impossibile* all the matter fields were removed from the world, is, I shall argue, the space-time analogue of the Cartesian 'ghost in the machine', to borrow Gilbert Ryle's famous pejorative phrase.

I should make it clear what this little book is not. It is not a textbook on relativity theory. It is not designed to teach the special or general theory. The latter only appears in the last chapter of the book, and there is even less about special relativistic dynamics in the sense of $E = mc^2$ and all that. What the book is about is the nature of special relativistic kinematics, its relation to space and time, and how it is supposed to fit in to general relativity. With the exception of the appendices, the book is designed to read—and here I borrow shamelessly from my other scientific hero, Charles Darwin—as one long argument.

Other research collaborators who have worked with me on topics related to the book are Katherine Brading, Peter Holland, Adolfo Maia, Roland Sypel, and Christopher Timpson; our interaction has been enjoyable and rewarding. I have benefitted a great deal from countless interactions with my Oxford colleagues Jeremy Butterfield and Simon Saunders; through their constructive criticism they have tried to keep me honest. I have also had very useful discussions on space-time matters and/or the history of relativity with Ron Anderson, Edward Anderson, Guido Bacciagaluppi, Yuri Balashov, Tim Budden, Marco Mamone Capria, Michael Dickson, Pedro Ferreira, Brendan Foster, Michel Ghins, Carl Hoefer, Richard Healey, Chris Isham, Michel Janssen, Oliver Johns, Clive Kilmister, Douglas Kutach, Nicholas Maxwell, Arthur Miller, John Norton, Hans Ohanian, Huw Price, Dragan Redžić, Rob Rynasiewicz, Graham Shore, Constantinos Skordis Richard Staley, Geoff Stedman, George Svetlichny, Roberto Torretti, Bill Unruh, David Wallace, Hans Westman, and Bill Williams. Antony Valentini suggested the first part of the title of this book, and has been a constant source of encouragement and inspiration. Useful references were kindly provided by Gordon Beloff and Michael Mackey. Katherine and Stephen Blundell volunteered to read the first draft of the book and apart from pointing out many typographical and spelling errors etc., made a number of important suggestions for improving clarity—and crucially encouraging noises. To all of these friends and colleagues I owe a debt of gratitude.

The two people who have had the greatest influence on my thinking about relativity are Julian Barbour and the late Jeeva Anandan who was also a collaborator. It is hard to summarize the multifarious nature of that influence, or to quantify the debt I owe them through their written work and many hours of conversation and contact. Julian Barbour taught me that the question 'what is motion?' is far deeper than I first imagined, and as a result made me entirely rethink the nature of space and time, and much else besides. Indeed, Julian's 1989 masterpiece on the history of dynamics *Absolute or Relative Motion?* came as a revelation to me; its combination of sure-footed history, conceptual insight and sheer exhilaration was unlike anything I had read before. I should perhaps clarify that my book is not designed to be a defence of a Leibnizian/Machian relational view of space-time of the kind Barbour has been articulating and defending with such brilliance in recent years, and in particular in his 1999 *The End of Time*. Although I have sympathies with this view, in my opinion the dynamical version of relativity theory is a separate issue and can be justified on much wider grounds, having essentially to do with good conceptual house-keeping. Jeeva Anandan, despite his exceptional abilities as a geometer, likewise drove home the lesson that physics is more than mathematics, and that operational considerations, though philosophically unfashionable, are essential in getting to grips with it.

I would also like to acknowledge the influence of the late Robert Weingard, whose enthusiasm for the subject of space-time rubbed off on me. He would almost certainly have found this book uncongenial in many ways, but his open-mindedness leads me to think, fondly, that he would not have dismissed it. I am indebted also to Jon Dorling and Michael Redhead, who in their different ways, taught me the ropes of philosophy of physics.

This book grew out of the experience of teaching a course over a number of years on the foundations of special relativity to second-year students in the Physics and Philosophy course at Oxford University. It is a privilege and pleasure to teach students of this calibre. I have gained a lot from their feedback through the years, and particularly that of Marcus Bremmer, James Orwell, Katrina Alexandraki, Michael Jampel, Hilary Greaves, and Eleanor Knox.

This project received vital prodding and cajoling from Peter Momtchiloff at Oxford University Press. His encouragement and faith are greatly appreciated. The comments, critical and otherwise, provided by the readers appointed by the Press to review the manuscript, Yuri Balashov, Carl Hoefer, and Steve Savitt, were very helpful and much appreciated. The copy-editor for the Press, Conan Nicholas, did a meticulous job on the original manuscript; I am very grateful to him for the resulting improvements. I thank Oliver Pooley and particularly Antony Eagle for patiently setting me straight about LaTeX. Abdullah Sowkaar D and Bhuvaneswari H Nagarajan at Newgen, India provided vital technical LaTeX-related help with the index, through the good services of Jason Pearce. Thanks go also to the staff of the Philosophy Library and the Radcliffe Science Library at Oxford University (physical sciences) for their ready and cheerful help.

Research related to different parts of this book was undertaken with the support of the Radcliffe Trust, the British Academy, and the Arts and Humanities Research Council (AHRB) of the UK. I am grateful to all these bodies, and to John Earman and Larry Sklar for their crucial help in securing the AHRB support.

Finally, without the sacrifices, patience, and love coming from my family—Maita, Frances, and Lucas—the book would never have seen the light of day. *Muitíssimo obrigado, meus queridos.*

Acknowledgements

Chapter 1 draws heavily on my 2003 paper 'Michelson, FitzGerald, and Lorentz: the origins of special relativity revisited', published in the *Bulletin de la Société des Sciences et des Lettres de Łódź*. Permission from the journal is gratefully acknowledged.

Some passages in Chapter 7 are taken verbatim from a paper entitled 'The Origins of the Spacetime Metric: Bell's Lorentzian pedagogy and its Significance in General Relativity', co-written with Oliver Pooley. It appeared in a volume published by Cambridge University Press in 2001. Some passages in Chapter 8 are likewise taken from a paper co-written in 2004 with Oliver Pooley, 'Minkowsi Space-time: a Glorious Non-entity' and yet to be published. I am grateful to both Dr Pooley and Cambridge University Press for permission to reproduce these passages.

Most of Appendix B is taken directly from a paper entitled 'Entanglement and Relativity', published in 2002 by the Department of Philosophy at the University of Bologna; its main author is Christopher Timpson. Permission from both is gratefully acknowledged.

Finally, thanks go to the Museum of the History of Science in Oxford, for permission to reproduce on the cover of this book the image of a late eighteenth-century waywiser in its possession.

H.R.B.

Contents

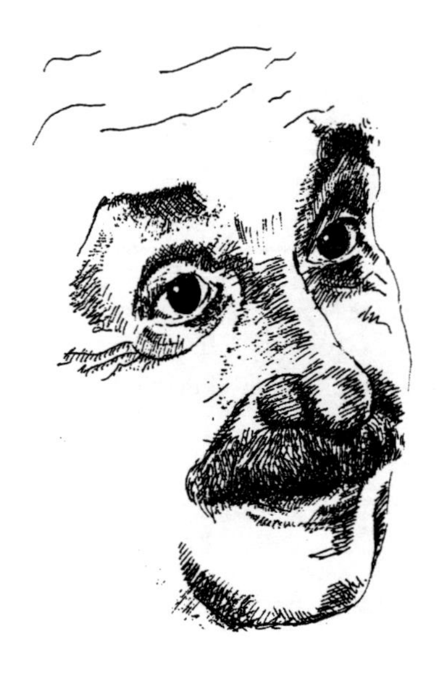

The “Great Relative.”

Name given Albert Einstein by Hopi Indians, 1921.

“The scientist finds his reward in what Henri Poincaré calls the joy of comprehension . . .”

Albert Einstein.

Pen drawing by the author.

1

Overview

The dogmas of the quiet past are inadequate to the stormy present.

Abraham Lincoln, 1862

1.1 WHEN THE WHOLE RIGMAROLE BEGAN

The claim that a particular theory in science had its true origins at this or that moment of time, in the emergence of this or that fundamental insight, is almost bound to be contentious. But there are developments, sometimes in the unpublished writings of a key figure, which deserve more recognition and fanfare in the literature for being truly seminal moments in the path to a given theory. In my opinion such a moment occurred in 1889. In the early part of that year George Francis FitzGerald, Professor of Natural and Experimental Philosophy at Trinity College Dublin, wrote a letter to the remarkable English auto-didact, Oliver Heaviside, concerning a result the latter had just obtained in the field of Maxwellian electrodynamics.[1] Heaviside had shown that the electric field surrounding a spherical distribution of charge should cease to have spherical symmetry once the charge is in motion relative to the ether. In this letter, FitzGerald asked whether Heaviside's distortion result might be applied to a theory of intermolecular forces. Some months later this idea would be exploited in a note by FitzGerald published in *Science*, concerning the baffling outcome of the 1887 ether-wind experiment of Michelson and Morley. FitzGerald's note is today quite famous, but it was virtually unknown until 1967. It is famous now because the central idea in it corresponds to what came to be known as the FitzGerald–Lorentz contraction hypothesis, or rather *to a distinct precursor of it*. The contraction effect is a cornerstone of the 'kinematic' component of the special theory of relativity proposed by Albert Einstein in 1905. But the FitzGerald–Lorentz explanation of the Michelson–Morley null result, known early on through the writings of Oliver Lodge, H. A. Lorentz, and Joseph Larmor, as well as through FitzGerald's relatively timid proposals to students and colleagues, was widely accepted as correct

[1] This chapter, which relies heavily on Brown (2003), is a brief outline of the main arguments of the book; references for all the works cited will be given in subsequent chapters.

before 1905. In fact it was accepted by the time of FitzGerald's untimely death in 1901 at the age of 49.

Following Einstein's brilliant 1905 work on the electrodynamics of moving bodies, and its geometrization by Minkowski which proved to be so important for the development of Einstein's general theory of relativity, it became standard to view the FitzGerald–Lorentz hypothesis as the right idea based on the wrong reasoning. I strongly doubt that this standard view is correct, and suspect that posterity will look kindly on the merits of the pre-Einsteinian, 'constructive' reasoning of FitzGerald, if not Lorentz. After all, even Einstein saw the limitations of his own approach based on the methodology of 'principle theories'. I need to emphasize from the outset, however, that I do not subscribe to the existence of the ether, nor recommend the use to which the notion is put in the writings of our two protagonists (which was very little). The merits of their approach have, as J. S. Bell stressed some years ago, a basis whose appreciation requires no commitment to the physicality of the ether.

There is, nonetheless, a subtle difference between the thinking of FitzGerald and that of Lorentz prior to 1905 that is of interest. What Bell called the 'Lorentzian pedagogy', and bravely defended, has, as a matter of historical fact, more to do with FitzGerald than Lorentz. Furthermore, the significance of Bell's work for general relativity has still not been fully appreciated.

1.2 FITZGERALD, MICHELSON, AND HEAVISIDE

A point charge at rest with respect to the ether produces, according to both intuition and Maxwell's equations, an electric field whose equipotential surfaces surrounding the charge are spherical. But what happens when the charge distribution is in uniform motion relative to the ether? Today, we ignore reference to the ether and simply exploit the Lorentz covariance of Maxwell's equations, and transform the stationary solution to one associated with a frame in relative uniform motion.

But in 1888, the covariance group of Maxwell's equations was yet to be discovered, let alone understood physically—the relativity principle not being thought to apply to electrodynamics—and the problem of moving sources required the solution of Maxwell's equations. These equations were taken to hold only relative to the rest frame of the ether. Oliver Heaviside found—it seems more on hunch than brute force—and published the solution: the electric field of the moving charge distribution undergoes a distortion, with the longitudinal components of the field being affected by the motion but the transverse ones not. The new equipotential surfaces define what came to be called a Heaviside ellipsoid.

The timing of Heaviside's distortion result was propitious, appearing as it did in the confused aftermath of the 1887 Michelson–Morley (MM) experiment. FitzGerald was one of Heaviside's correspondents and supporters, and found, like all competent ether theorists, the null result of this fantastically sensitive experiment a mystery. Null results in earlier *first-order* ether wind experiments had all been explained in terms of the Fresnel drag coefficient, which would in 1892 receive an electrodynamical underpinning of sorts in the work of Lorentz. But by early 1889 no one had accounted for the absence of noticeable fringe shifts in the second-order MM experiment. How could the apparent isotropy of the two-way light speed inside the Michelson interferemeter be reconciled with the seeming fact that the laboratory was speeding through the ether? Why didn't the ether wind blowing through the laboratory manifest itself when the interferometer was rotated?

The conundrum of the MM null-result was surely in the back of FitzGerald's mind when he made an intriguing suggestion in that letter to Heaviside in January 1889. The suggestion was simply that a Heaviside distortion might be applied 'to a theory of the forces between molecules' of a rigid body. FitzGerald had no more reason than anyone else in 1889 to believe that these intermolecular forces were electromagnetic in origin. No one knew. But if these forces too were rendered anisotropic by the mere motion of the molecules, which FitzGerald regarded as plausible in the light of Heaviside's work, then the shape of a rigid body would be altered as a consequence of the motion. This line of reasoning was briefly spelt out, although with no *explicit* reference to Heaviside's work, in a note that FitzGerald published later in the year in the American journal *Science*. This was the first correct insight into the mystery of the MM experiment when applied to the stone block on which the Michelson interferometer was mounted. But the note sank into oblivion; FitzGerald did not bother to confirm that it was published, and seems never to have referred to it, though he did promote his deformation idea in lectures, discussions, and correspondence. His relief when he discovered that Lorentz was defending essentially the same idea was palpable in a good-humoured letter he wrote to the great Dutch physicist in 1894, which mentioned that he had been 'rather laughed at for my view over here'.

It should be noted that FitzGerald never seems to have used the words 'contraction' or 'shortening' in connection with the proposed motion-induced change of the body. The probable reason is that he did *not* have the purely longitudinal contraction, now ubiquitously associated with the 'FitzGerald–Lorentz hypothesis', in mind. It is straightforward to show, though not always appreciated, that the MM result does not demand it. Any deformation (including expansion) in which the ratio of the suitably defined transverse and longitudinal length change factors equals the Lorentz factor $\gamma = (1 - v^2/c^2)^{-1/2}$ will do, and there are good reasons to think that this is what FitzGerald meant, despite some claims to the contrary on the part of historians. It is certainly what Lorentz had in mind for several years after 1892, when he independently sought to account for the MM result by appeal to a change in the dimensions of rigid bodies when put into motion.

1.3 EINSTEIN

In his masterful review of relativity theory of 1921, Wolfgang Pauli was struck by the difference between Einstein's derivation and interpretation of the Lorentz transformations in his 1905 paper and that of Lorentz in his theory of the electron. Einstein's discussion, noted Pauli, was in particular 'free of any special assumptions about the constitution of matter', in strong contrast with Lorentz's treatment. He went on to ask: 'Should one, then, completely abandon any attempt to explain the Lorentz contraction atomistically?'

It may surprise some readers to learn that Pauli's answer was negative. Be that as it may, it is a question that deserves careful attention, and one that, if not haunting him, then certainly gave Einstein unease in the years that followed the full development of his theory of relativity.

Einstein realized, possibly from the beginning, that the first, 'kinematic' section of his 1905 paper was problematic, that it effectively rested on a false dichotomy. What is kinematics? In the present context it is the universal behaviour of rods and clocks in motion, as determined by the inertial coordinate transformations. And what are rods and clocks, if not, in Einstein's own later words, 'moving atomic configurations'? They are macroscopic objects made of micro-constituents—atoms and molecules—held together largely by electromagnetic forces. But it was the second, 'dynamical' section of the 1905 paper that dealt with the covariant treatment of Maxwellian electrodynamics. Einstein knew that the first section was not wholly independent of the second, and in 1949 would admit that the treatment of rods and clocks in the first section as primitive, or 'self-sustained' entities was a 'sin'. The issue is essentially the same one that Pauli had stressed in 1921:

> The contraction of a measuring rod is not an elementary but a very complicated process. It would not take place except for the covariance with respect to the Lorentz group of the basic equations of electron theory, as well as those laws, as yet unknown to us, which determine the cohesion of the electron itself.

Pauli is here putting his finger on two important points: that the distinction between kinematics and dynamics is not fundamental, and that to give a full treatment of the dynamics of length contraction was still beyond the resources available in 1921, let alone 1905. And this latter point was precisely the basis of the excuse Einstein later gave for his 'principle theory' approach—modelled on thermodynamics—in 1905 in establishing the Lorentz transformations.

The singular nature of Einstein's argumentation in the kinematical section of his paper, its limitations and the recognition of these limitations by Einstein himself, will be discussed in detail below. It is argued that there is in fact a significant dynamical element in Einstein's reasoning in that section, specifically in relation to the use of the relativity principle, and that it is unclear whether Einstein himself appreciated this. The main lesson that emerges, as I see it, is that the special theory of relativity is incomplete without the assumption that the quantum theory of *each* of the fundamental non-gravitational interactions—and not just electrodynamics—is Lorentz-covariant. This lesson was anticipated as early as 1912 by W. Swann, and established in a number of his papers up to 1941. It was independently advocated by L. Jánossy in 1971, and reinforced in the didactic approach to special relativity advocated by J. S. Bell in 1976, to which we return shortly.

Swann's unsung achievement was in effect to spell out in detail the meaning of Pauli's 1921 warning above. His incisive point was that the Lorentz covariance of Maxwellian electrodynamics, for example, has no clear connection with the claim that electrodynamics satisfies the relativity principle, unless it could be established that the Lorentz transformations are more than just a formal change of variables and actually codify the behaviour of moving rods and clocks. But the validity of this last assumption depends on our best theory of the micro-constitution of stable macroscopic objects. Or rather, it depends on a fragment of quantum theory (for it could not be other than a quantum theory): that at the most fundamental level all the interactions involved in the composition of matter, whatever their nature, are Lorentz covariant. It must have been galling for Einstein to recognize this point, given his lifelong struggle with the quantum. It is noteworthy that although he repeats in his 1949 *Autobiographical Notes* the imperative to understand rods and clocks as structured, composite bodies, which he had voiced as early as 1921, he makes no concession to the great strides that had been made in the quantum theory of matter in the intervening years.

1.4 FITZGERALD AND BELL'S 'LORENTZIAN PEDAGOGY'

In 1999, Oliver Pooley and I referred to this insistence on this role of quantum theory in special relativity as the 'truncated' version of the 'Lorentzian pedagogy' advocated by J. S. Bell in 1976. The *full* version of this pedagogy involves providing a constructive model of the matter making up a rod and/or clock and solving the equations of motion in the model. Bell's terminology is slightly misplaced: it would be more appropriate still to call this reasoning the 'FitzGeraldian pedagogy'!

Bell's model (which is discussed at greater length below) has as its starting point a single atom built of an electron circling a much more massive nucleus. Using not much more than Maxwellian electrodynamics (taken as valid relative to the rest frame of the nucleus), Bell determined that the orbit undergoes the familiar relativistic longitudinal contraction, and its period changes by the familiar 'Larmor' dilation. Bell claimed that a rigid arrangement of such atoms as a whole would do likewise, given the electromagnetic nature of the interatomic/molecular forces. He went on to demonstrate that there is a system of primed variables such that the description of the *uniformly* moving atom with respect to them is the same as the description of the stationary atom relative to the orginal variables—and that the associated transformations of coordinates are precisely the familiar Lorentz transformations. But it is important to note that Bell's prediction of length contraction and time dilation is based on an analysis of the field surrounding a (gently) *accelerating* nucleus and its effect on the electron orbit. The significance of this point will become clearer in the next section.

Bell cannot be berated for failing to use a truly satisfactory model of the atom; he was perfectly aware that his atom is unstable and that ultimately only a quantum theory of both nuclear and atomic cohesion would do. His aim was primarily didactic. He was concerned with showing us that

> [W]e need not accept Lorentz's philosophy [of the reality of the ether] to accept a Lorentzian pedagogy. Its special merit is to drive home the lesson that the laws of physics in any *one* reference frame account for all physical phenomena, including the observations of moving observers.

For Bell, it was important to be able to demonstrate that length contraction and time dilation can be derived independently of coordinate transformations—independently of a technique involving a change of variables.

But this is not strictly what Lorentz had done in his treatment of moving bodies, despite Bell's claim that he followed very much Lorentz's approach. (It is noteworthy both that Bell gives no references to Lorentz's papers, and admits that the inspiration for the method of integrating equations of motion in a model of the sort he presented was 'perhaps' a remark of Larmor.)

The difference between Bell's treatment and Lorentz's theorem of corresponding states that I wish to highlight is not that Lorentz never discussed accelerating systems. He didn't, but of more relevance is the point that Lorentz's treatment, to put it crudely, is (almost) mathematically the modern change-of-variables-based-on-covariance approach but with the wrong physical interpretation. Lorentz used auxiliary coordinates, field strengths, and charge and current densities associated with an observer co-moving with the laboratory, to set up states of the physical bodies and fields that 'correspond' to states of these systems when the laboratory is at rest relative to the ether, both being solutions of Maxwell's equations. Essentially, prior to Einstein's work, Lorentz failed to understand (even when Poincaré pointed it out) that the auxiliary quantities were precisely the quantities that the co-moving observer would be measuring, and not mere mathematical devices. But then to make contact with the actual physics of the ether-wind experiments, Lorentz needed to make a number of further complicating assumptions, the nature of which we return to later. Suffice it to say here that the whole procedure was limited in practice to stationary situations associated with optics, electrostatics, and magnetostatics.

The upshot was an explanation of the null results of the ether-wind experiments that was if anything mathematically simpler, but certainly conceptually much more complicated—not to say obscure—than the kind of exercise Bell was involved with in his 1976 essay. It cannot be denied that Lorentz's argumentation, as Pauli noted in comparing it with Einstein's, is dynamical in nature. But Bell's procedure for accounting for length contraction is in fact much closer to FitzGerald's 1889 thinking based on the Heaviside result, summarized in section 1.2 above. In fact it is essentially a generalization of that thinking to the case of accelerating bodies.

Finally, a word about time dilation. It was seen above that Bell attributed its discovery to Joseph Larmor, who indeed had partially—very partially—understood the phenomenon in his 1900 *Aether and Matter*, a text based on papers Larmor had published in the very last years of the nineteenth century. It is still widely believed that Lorentz failed to anticipate time dilation before the work of Einstein in 1905, as a consequence of failing to see that the 'local' time appearing in his own (second-order) theorem of corresponding states was more than just a mathematical artifice, but rather the time as read by suitably synchronized clocks at rest in the moving system. It is interesting that if one does an analysis of the famous variation of the MM experiment performed by Kennedy and Thorndike in 1932, exactly in the spirit of Lorentz's 1895 analysis of the MM experiment and *with no allowance for time dilation*, then the result, taking into account the original MM outcome too, is *the wrong kind of deformation for moving bodies*.[2] It can easily be shown that rods must contract transversely by the factor γ^{-1} and longitudinally by the factor γ^{-2}. One might be tempted to conclude that Lorentz, who had opted for purely longitudinal contraction (for dubious reasons), was lucky that it took so long for the Kennedy–Thorndike experiment to be performed!

But the conclusion is probably erroneous. In 1899, as Michel Janssen recently spotted, Lorentz had already discussed yet another interesting variation of the MM experiment, suggested a year earlier by the French physicist A. Liénard, in which transparent media were placed in the arms of the interferometer. The experiment had not been performed, but Lorentz both suspected that a null result would still be obtained, and realized that shape deformation of the kind he and FitzGerald had proposed would not be enough to account for it. What was lacking, according to Lorentz? Amongst other things, the claim that the frequency of oscillating electrons in the light source is lower in systems in motion than in systems at rest relative to the ether. Lorentz had pretty much the same (limited) insight into the nature of time dilation as Larmor did, at almost the same time. It seems that the question of the authorship of time dilation is ripe for reanalysis, and we return to this issue in Chapter 4.

[2] Kennedy and Thorndike have as the title of their paper 'Experimental Establishment of the Relativity of Time', but their experiment does not imply the existence of time dilation unless it is assumed that motion-induced deformation in rigid bodies is purely longitudinal—indeed, just the usual length contraction. As mentioned above, this specific kind of deformation is not a consequence of the MM experiment, and was still not established experimentally in 1932 (although it was widely accepted). What the Kennedy–Thorndike experiment established unequivocally, in conjunction with the MM experiment, is that the two-way light speed is (inertial) frame-independent.

1.5 WHAT SPACE-TIME IS NOT

If you visit the Museum of the History of Science in Oxford, you will find a number of fine examples of eighteenth- and nineteenth-century devices called *waywisers*, designed to measure distances along roads. Typically, these devices consist of an iron-rimmed wheel, connected to a handle and readout dial. The dial registers the number of revolutions of the wheel as the whole device is pulled along the road, and has hands which indicate yards and furlongs/miles. (Smaller versions of the waywiser are seen being operated by road maintenance crews today in the UK, and are sometimes called measuring wheels.) The makers of these original waywisers had a premonition of relativity! For the dials on the waywisers typically look like clocks. And true, ideal clocks are of course the waywisers, or hodometers, of time-like paths in Minkowski space-time.

The *mechanism* of the old waywiser is obvious; there is no mystery as to how friction with the road causes the wheel to revolve, and how the information about the number of such 'ticks' is mechanically transmitted to the dial. But the true clock is more subtle. There is no friction with space-time, no analogous mechanism by which the clock reads off four-dimensional distance. How does it work?

One of Bell's professed aims in his 1976 paper on 'How to teach relativity' was to fend off 'premature philosophizing about space and time'. He hoped to achieve this by demonstrating with an appropriate model that a moving rod contracts, and a moving clock dilates, *because of how it is made up and not because of the nature of its spatio-temporal environment.* Bell was surely right. Indeed, if it is the structure of the background spacetime that accounts for the phenomenon, by what mechanism is the rod or clock informed as to what this structure is? How does this material object get to know which type of space-time—Galilean or Minkowskian, say—it is immersed in?

Some critics of Bell's position may be tempted to appeal to the general theory of relativity as supplying the answer. After all, in this theory the metric field *is* a dynamical agent, both acting on, and being acted upon by, the presence of matter. But general relativity does not come to the rescue in this way (and even if it did, the answer would leave special relativity looking incomplete). Indeed the Bell–Jánossy–Pauli–Swann lesson—which might be called the *dynamical* lesson—serves rather to highlight a feature of general relativity that has received far too little attention to date. It is that in the absence of the strong equivalence principle, the metric $g_{\mu\nu}$ in general relativity has no automatic *chronometric* operational interpretation.

For consider Einstein's field equations

$$R_{\mu\nu} - \frac{1}{2}g_{\mu\nu}R = 8\pi GT_{\mu\nu} \tag{1.1}$$

where $R_{\mu\nu}$ is the Ricci tensor, R the curvature scalar, $T_{\mu\nu}$ the stress energy tensor associated with matter fields, and G the gravitational constant. A possible space-time, or metric field, corresponds to a solution of this equation, but nothing in the form of the equation determines either the metric's signature or its operational significance. In respect of the last point, the situation is not wholly dissimilar from that in Maxwellian electrodynamics, in the absence of the Lorentz force law. In both cases, the ingredient needed for a direct operational interpretation of the fundamental fields is missing.

But of course there is more to general relativity than the field equations. There is, besides the specification of the Lorentzian signature for $g_{\mu\nu}$, the crucial assumption that locally physics looks Minkowskian. (Mathematically of course the tangent spaces are automatically Minkowskian, but the issue is one of physics, not mathematics.) It is a component of the strong equivalence principle that in 'small enough' regions of space-time, for most practical purposes the physics of the non-gravitational interactions takes its usual Lorentz covariant form. In short, as viewed from the perspective of the local freely falling frames, special relativity holds when the effects of space-time curvature—tidal forces—can be ignored. It is this extra assumption, which brings in *quantum* physics even if this point is rarely emphasized, that guarantees that ideal clocks, for example, can both be defined and shown to survey the postulated metric field $g_{\mu\nu}$ when they are moving inertially. Only now is the notion of proper time linked to the metric. But yet more has to be assumed before the metric gains its full, familiar chronometric significance.

The final ingredient is the so-called *clock hypothesis* (and its analogue for rods). This is the claim that when a clock is accelerating, the effect of motion on the rate of the clock is no more than that associated with its instantaneous velocity—the acceleration adds nothing. This allows for the identification of the integration of the metric along an *arbitrary* time-like curve—not just a geodesic—with the proper time. This hypothesis is no less required in general relativity than it is in the special theory. The justification of the hypothesis inevitably brings in dynamical considerations, in which forces internal and external to the clock (rod) have to be compared. Once again, such considerations ultimately depend on the quantum theory of the fundamental non-gravitational interactions involved in material structure.

In conclusion, the operational meaning of the metric is ultimately made possible by appeal to quantum theory, in general relativity as much as in the special theory. The only, and significant, difference is that in special relativity, the Minkowskian metric is no more than a codification of the behaviour of rods and clocks, or equivalently, it is no more than the Kleinian geometry associated with the symmetry group of the quantum physics of the non-gravitational interactions in the theory of matter. In general relativity, on the other hand, the $g_{\mu\nu}$ field is an autonomous dynamical player, physically significant even in the absence of the usual 'matter' fields. But its meaning as a carrier of the *physical* metrical relations between space-time points is a bonus, the gift of the strong equivalence principle and the clock (and rod) hypothesis. The problem in general relativity is that the matter fields responsible for the stress-energy tensor appearing in the field equations are classical, and thus there is a deep-seated tension in the story about how the metric field gains its chronometric operational status.

1.6 FINAL REMARKS

It seems to be widely accepted today that Einstein owed little to the Michelson–Morley experiment in his development of relativity theory. Yet the null result cannot but have buttressed his conviction in the validity of the relativity principle, or at least its applicability to electromagnetic phenomena. And as we shall see later, in 1908 Einstein wrote to Sommerfeld clarifying the methodological analogy between his 1905 relativity theory and classical thermodynamics. It was clear here (and elsewhere in Einstein's writings) that by stressing this connection with thermodynamics Einstein was stressing the *limitations* of his theory rather than its strengths—and his explicit point was that even 'half' a solution is better than none to the dilemma posed by the Michelson–Morley result.

Be that as it may, there is no doubt about the spur the MM experiment gave to the insights gained by FitzGerald and Lorentz concerning the effects of motion on the dimensions of rigid bodies. It is my hope that commentators in the future will increasingly recognize the importance of these insights, and that the contributions of the two pioneers will emerge from the shadow cast by Einstein's 1905 'kinematic' analysis. As Bell argued, the point is not that Einstein erred, so much as that the messier, less economical reasoning based on 'special assumptions about the composition of matter' can lead to greater insight, in the manner that statistical mechanics can offer more insight than thermodynamics. The longer road, Bell reminded us, may lead to more familiarity with the country.

2

The Physics of Coordinate Transformations

> It [the law of inertia] reads in detailed formulation necessarily as follows: Matter points that are sufficiently separated from each other move uniformly in a straight line—provided that the motion is related to a suitably moving coordinate system and that the time is suitably defined. Who does not feel the painfulness of such a formulation? But omitting the postscript would imply a dishonesty.
>
> Albert Einstein[1]

> The first law . . . is a logician's nightmare . . . To teach Newton's laws so that we prompt no questions of substance is to be unfaithful to the discipline itself.
>
> J. S. Rigden[2]

2.1 SPACE-TIME AND ITS COORDINATIZATION

It is common in discussions of the principle of general covariance in Einstein's general theory of relativity to find the claim that coordinates assigned to events are merely labels. Since physics, or the objective landscape of events, cannot depend on the labelling systems we choose to distinguish events, it would seem to follow that in their most fundamental form the laws of physics should be coordinate-general, or 'generally covariant' as it is usually put.

Discerning students should be puzzled on a number of grounds. (A) Before we consider labelling them, what *physically* distinguishes two different events of exactly the same kind? (B) Why doesn't this labelling argument apply to all theories, and not just general relativity? (C) And how is it that coordinate transformations—which are presumably nothing more than re-labelling schemes—can in some cases contain physics? Indeed, if contrary to the normal procedure, we were to learn general relativity prior to special relativity, wouldn't we be puzzled to see apparently physical effects such as length contraction and time dilation emerge from the Lorentz transformations between local inertial coordinate systems?

[1] Einstein (1920); English translation in Pfister (2004). [2] Rigden (1987).

In his *General Relativity from A to B*, Robert Geroch gives us the following account of the notion of 'event': 'By an event we mean an idealized occurrence in the physical world having extension in neither space nor time. For example, "the explosion of a firecracker" or "the snapping of one's fingers" would represent an event.'[3]

Geroch is careful of course to qualify the firecracker as 'very small, very fast-burning'—after all, events are supposed to be points in an appropriate space. What is important for our present purposes is Geroch's account of the sameness of events: 'We regard two events as being "the same" if they coincide, that is, if they "occur at the same place at the same time." That is to say, we are not now concerned with how an event is marked—by firecracker or finger-snap—but only with the thing itself.'[4]

Geroch's intuition is clearly that there is a difference between the localized material thing that 'marks' the event and the event in itself. This kind of view—modulo terminological variations—has a prestigious lineage. For example, Minkowski made a distinction in his famous Cologne lecture of 1908 between 'substantial' and 'world' points:

> I still respect the dogma that space and time have independent existence. A point of space at a point of time, that is a system of values x, y, z, t, I will call a *world-point*. The multiplicity of all thinkable x, y, z, t systems of values we will christen the *world*. ... Not to leave a yawning void anywhere we will imagine that everywhere and everywhen there is something perceptible. To avoid saying 'matter' or 'electricity' I will use for this something the word 'substance'. We fix our attention on the substantial point which is at the world point x, y, z, t, and imagine we are able to recognize this substantial point at any other time.[5]

Einstein made similar remarks prior to 1915. But whereas Minkowski and the early Einstein were in no doubt as to the reality of the points underpinning the material markers, this cannot be said for Geroch. The somewhat shadowy nature of the world-point or event (in Geroch's strict sense of the word) prompts him to to raise and then explicitly avoid questions as to its reality, particularly after wondering whether an event is not better characterized as an 'idealized *potential* occurrence ...'. He ends up sidestepping the reality issue by claiming that 'Relationships between events—that is what we are after.'

Geroch is right to be cagey about the reality of the underlying events as he defines them. The usual appeal to the existence of the physical continuum, or 'manifold' of *featureless* space-time points ineluctably raises conceptual problems that were the backdrop to the great debate in the late seventeenth and early eighteenth centuries between Newton and Leibniz on the nature of space and time. The main such problem goes as follows.

It is only the markers, to use Geroch's terminology, and not the events proper that come under our senses. We could imagine two universes with identical

[3] Geroch (1978), p. 3. [4] op. cit., p. 4. [5] Minkowski (1909).

arrangements of markers, and identical systems of observable relations between them, which differ only by the way the markers are related to the space-time points. Thus, Leibniz considered two material universes differing only by the locations in space God decides to put them. Einstein, roughly two centuries later, found himself likewise considering two empirically indistinguishable space-times that differ only by the way the metric tensor field $g_{\mu\nu}$ in each relates to the background 4-dimensional point manifold. Now Leibniz famously dismissed his pair of cosmological alternatives as a fancy, on the grounds that it violated both the Principle of Identity of Indiscernibles and the Principle of Sufficient Reason. The two universes were for Leibniz but one and the same thing. God could thus avoid the hopeless task of rationally deciding where to put the universe in space because space is not a separate thing! In 1915, Einstein, in tackling what was later called the Hole Problem, came to reject the reality of the space-time manifold essentially on the grounds that such a position allowed his gravitational field equations to avoid the spectre of underdetermination—the analogue of Leibniz's spectre of divine indecision.[6] The way Einstein put it in 1952 was: space-time is a 'structural quality of the field', not the other way around.

I think there are indeed good grounds for questioning the existence of the physical space-time manifold, or the set of events in the strict sense of Geroch, at least if the manifold points have no distinguishing features. (We return to this issue in Chapter 9.) But if the space-time points as they are usually understood do not exist, it is not entirely clear why we, like Geroch, ought to concern ourselves with relationships between them. On the other hand, Geroch is surely right to think that the marker is not enough to get hold of the notion of a space-time point.

Recall we can think of the markers as suitably idealized explosions, collisions of point-particles, flashes of light and so on—the kinds of things physicists typically mention when asked to provide examples of 'events'. We might even try to be more technical and insist on characterizing a marker as the set of values at a point of the (components of the) most fundamental fields in our best physical theories. Whatever your favourite example of a marker is, it is bound to occur at many distinct points in space and at many times in the history of the universe. The very existence of a lawlike structure in the universe, of the fact that physics deals with empirical *regularities*, makes this virtually inevitable. The flash of light, or the collision of particles or whatever, taken in its idealized, pristine, localized sense, is *not a one-off*. (Something like this point is suggested at the end of the quote from Minkowski above.) As a consequence, in the language of the mathematical physicist, there simply cannot be a one-to-one correspondence between the set

[6] It might more correctly be said that when Einstein realized that the 'diffeomorphically related' spacetimes were physically indisinguishable, he ceased to believe them physically (as opposed to mathematically) distinct, thus adopting a stance with clear echoes of Leibniz. Given this stance, the apparent underdetermination of the field equations that is a consequence of their general covariance evaporates.

of distinct marks and the set of points that is the space-time manifold. Now this may not be considered a problem if the coordinates are used to label, first and foremost, the manifold points. But this is both conceptually questionable (as we have seen) and operationally mysterious. If coordinate systems are labelling schemes we impose on the world, how is that we go about coordinatizing space-time points which we do not and cannot see?

Let's consider what it is that distinguishes two flashes of light that, in Geroch's terms, do not coincide. The distinction does not lie in the fact that they have different coordinates. They have different coordinates because they are distinct, and they are distinct not in virtue of what they are *locally* but in virtue of the fact that *they stand in different relations to the rest of the universe*—to the rest of the markers. It is because those relations are in principle discernible that we can say that the same markers can occur at different space-time points. So rather than think of a space-time point, i.e. an event in Geroch's strict sense, as a self-contained localized atom of the invisible uniform space-time manifold, we might more usefully think of it as the view of the universe from a point.

This is how Julian Barbour put the idea in 1982.

> Minkowski, Einstein, and Weyl invite us to take a microscopic look, as it were, for little featureless grains of sand, which, closely packed, make up space-time. But Leibniz and Mach suggest that if we want to get a true idea of what a point of space-time is like we should look *outward* at the universe, not inward into some supposed amorphous treacle called the space-time manifold. The complete notion of a point of space-time in fact consists of *the appearance of the entire universe as seen from that point.* Copernicus did not convince people that the earth was moving by getting them to examine the earth but rather the heavens. Similarly, the reality of different points of space-time rests ultimately on the existence of different (coherently related) viewpoints of the universe as a whole. Modern theoretical physics will have us believe that the points of space are uniform and featureless; in reality, they are incredibly varied, as varied as the universe itself.[7]

The preceding discussion represents an attempt to briefly address question (A) above. Question (B) will be left until our discussion of general relativity. The rest of this chapter is designed to address various issues raised by question (C).

2.2 INERTIAL COORDINATE SYSTEMS

Inertia, before Einstein's general theory of relativity, was a miracle. I do not mean the existence of inertial mass, but the postulate that force-free (henceforth *free*) bodies conspire to move in straight lines at uniform speeds while being unable,

[7] Barbour (1982). A discussion of the meaning of points in space that is similar in spirit is found in Poincaré (1952), section 8 of chap. V, pp. 84–8. It should be noted that, from the point of view of quantum theory, the familiar space-time events we have been discussing are only 'effective' notions specifically the outcome of quantum decoherence.

by *fiat*, to communicate with each other. It is probably fair to say that anyone who is not amazed by this conspiracy has not understood it. (And the coin has two sides: anyone who is not struck by the manner in which the general theory is able to explain the conspiracy—a feature of his own theory to which Einstein was oblivious in 1915—has failed to appreciate its strength, as we shall see later.)

Newton's first law of motion, for that is what we have been discussing, can be construed as an existence claim. Inertial coordinate systems are those special coordinate systems relative to which the above conspiracy, involving rectilinear uniform motions, unfolds. *Qua* definition, this statement has of course no content. What *has* content is the claim that such a coordinate system exists, applicable to *all* the free bodies in the universe.[8] Needless to say, it would be nice to give a coordinate-independent formulation of the same principle, and we shall return to that shortly. Right now we need to clarify what role Newton's first law plays in the special theory of relativity.[9]

The special theory of 1905, together with its refinements over the following years, is, in one important respect, *not* the same theory that is said to be the restriction of the general theory in the limit of zero gravitation (i.e. zero tidal forces, or space-time curvature). The nature of this limiting theory, and its ambiguities, will be discussed later; for our present purposes we shall associate it with the local, tangent-space structure of GR, which to a good approximation describes goings-on in sufficiently 'small' regions of space-time. But in this picture, local inertial coordinate systems are freely falling systems. They are not in Einstein's 1905 theory. Einstein stated explicitly in his 1905 paper that the inertial coordinate systems were the ones in which Newton's laws held good, by which he really meant the first law, and of course for Newton a freely falling object is accelerating with respect to inertial frames—it is not free. For the moment, we will follow the 1905, not the 1915, Einstein.

2.2.1 Free Particles

There is little doubt that Newton's first law (inspired by Descartes' 1644 principle of inertial motion) is empty unless one can demarcate between force-free bodies and the rest. Precisely how this demarcation is to be understood is still a moot point. In his Definition IV of the *Principia*, Newton states that a force is essentially an agency that causes bodies to deviate from their natural inertial motions; this

[8] See e.g., Weyl (1952), p. 178, and Bergmann (1976), p. 8. This view of the first law as an existence claim has been criticized by Earman and Friedman (1973), who claim that it 'is not empirical in the way the second law is; rather it is an [unsuccessful] attempt to specify part of the structure of Newtonian space-time' (p. 337). I fail to see why the existence claim is not empirical.

[9] An unorthodox approach to inertia in Newtonian mechanics (or an important sector thereof) emerges from the application of a global 'best-matching' procedure developed by Barbour and Bertotti (1982). Whether this approach, inspired by the relationism of Leibniz, Mach, and Poincaré, makes inertia less miraculous is a moot point, but it notably establishes a deep connection with inertial mass, see Pooley and Brown (2002), p. 199.

hardly helps. But implicit in Newton's mechanics is the assumption that forces on a given body are caused by the presence of other bodies, whether acting by contact or by action-at-a-distance as in the case of gravity. If all forces of the latter variety likewise fall off sufficiently quickly with distance, then bodies sufficiently far from all other bodies are effectively free. There is little question too that Newton would have accepted that bodies constrained to move on flat friction-free surfaces by forces orthogonal to the surface would move inertially.

In a recent careful treatment of Newton's first law, Herbert Pfister prefers to avoid the approximate and even 'logically fallacious' nature of Newton's treatment,[10] in favour of defining free particles as inactive test objects with only one essential physical property: mass. Thus they should have zero charge, magnetic moment, higher electromagnetic multipole moments, intrinsic angular momentum, all higher mass multipole moments; '(nearly) any other physical property imaginable, or for which experimentalists have invented a measuring device should also be zero'.[11] I cannot help wondering if such an account does not rely too much on hindsight—whether indeed the definition of the properties that are supposed to have a null value does not ultimately refer to the very inertial frames we are trying to construct. My purpose here is not so much to resolve the issue—though my sympathies here are more with Newton than Pfister—but rather to stress that it is an important one.[12]

2.2.2 Inertial Coordinates

For the moment, let us assume that point particle paths are 1-dimensional submanifolds defined within a 4-dimensional space-time manifold M. In a given coordinate system x^μ, ($\mu = 0, \ldots, 3$) suppose that the path of any free particle can be expressed thus

$$d^2x^\mu/d\tau^2 = 0, \tag{2.1}$$

where τ is a monotonic parameter on the path $x^\mu(\tau)$ in question. Integration of (2.1) yields

$$x^\mu(\tau) = x^\mu(0) + \tau v^\mu(0), \tag{2.2}$$

[10] Pfister (2004) states, incorrectly in my view, that the definition of 'force' and 'force-free' is reserved to Newton's second law, and furthermore that since the gravitational and electromagnetic forces have an infinitely long range, it is 'impossible in any practical case to say what distance to other objects is "big enough" . . .'. I doubt such pessimism is truly warranted.

[11] Pfister (2004), p. 54.

[12] In another careful treatment of Newton's first law, J. L. Anderson seems to regard it as part of the *definition* of free bodies that they move in straight lines in space-time, and that the non-trivial existence claim associated with the first law concerns just 'the ensemble of straight lines that form part of the geometric structure of Newtonian space-time' (Anderson (1990), p. 1193). What is unclear in this account is what the physical objects are that trace out the straight lines.

where $v^\mu(0) = \mathrm{d}x^\mu/\mathrm{d}\tau$ at $\tau = 0$, so we obtain a straight line in the 4-dimensional manifold. Yet this simple description is of course coordinate-dependent. Imagine an arbitrary (not necessarily 'projective', or even linear) coordinate transformation[13] $x^\mu \longrightarrow x'^\mu(x^\nu)$, along with an arbitrary parameter transformation $\tau \longrightarrow \lambda(\tau)$. Then (2.1) is transformed into

$$\frac{\mathrm{d}^2x'^\mu}{\mathrm{d}\lambda^2} + \Pi^\mu{}_{\nu\sigma}(x'^\rho)\frac{\mathrm{d}x'^\nu}{\mathrm{d}\lambda}\frac{\mathrm{d}x'^\sigma}{\mathrm{d}\lambda} = \frac{\mathrm{d}^2\tau}{\mathrm{d}\lambda^2}\frac{\mathrm{d}\lambda}{\mathrm{d}\tau}\frac{\mathrm{d}x'^\mu}{\mathrm{d}\lambda}, \tag{2.3}$$

where

$$\Pi^\mu{}_{\nu\sigma} = -\frac{\partial^2 x'^\mu}{\partial x^\rho \partial x^\gamma}\frac{\partial x^\rho}{\partial x'^\nu}\frac{\partial x^\gamma}{\partial x^\sigma}. \tag{2.4}$$

(Note that from here on we use the Einstein convention for repeated indices.) It may be 'painful' to see how easily the simple form of (2.1) is lost (see the quotation from Einstein in the epigraph at the beginning of this chapter), but let us not lose sight of the main point. A kind of highly non-trivial pre-established harmony is being postulated, and it takes the form of the claim that there exists a coordinate system x^μ and parameters τ such that (2.1) holds for each and every free particle in the universe. Now we are not yet at the principle of inertia as standardly construed, but a word here about geometry.

Hermann Weyl was the first to notice that the structure we have just defined is that of a projective geometry,[14] and the point was given further prominence in the famous 1972 paper of Ehlers, Pirani, and Schild[15] on the operational meaning of the geometrical structures of space-time. Pfister has stressed that straight lines can be characterized in projective geometry in a coordinate independent way: he defines them as paths that fulfill the so-called Desargues property. (*Desargues' theorem* is the statement that if corresponding sides of two triangles meet in three points lying on a straight line, then corresponding vertices lie on three concurrent lines.) We are to suppose then that free particles follow paths which are straight in this sense.

> In detail, the first four paths of such a construction define an inertial system. That, however, all other free particles also move on straight lines with respect to this inertial system and do so independently of the mass and many other inner properties of the particles, belongs to the most fundamental and marvellous facts of nature.[16]

Now it is a remarkable property of Desargues' theorem that it is *self-dual*. If we interchange the parts played by the words 'points' and 'lines', the new proposition

[13] Projective coordinate transformations are defined in section 3.3 below.

[14] Projective geometry deals with properties and invariants of geometric figures under projection. It is based on the notions of collinearity and concurrence, and so its concerns are with straight lines and points; the notions of distance, angle, and parallelism are absent.

[15] Ehlers *et al.* (1972); see also Ehlers (1973), §2.4. [16] Pfister (2004).

is equivalent to the old. Indeed, all the propositions in projective geometry occur in dual pairs such that given either proposition of the pair, the other can be inferred by this interchange. It is unclear to me whether the material particles in the Newtonian picture break this symmetry, and part of the uncertainty surrounding the matter has to do with the fact that in the 4-dimensional projective space under consideration no clear distinction between space and time has been elucidated.

Both Descartes and Newton were clearly saying something stronger than the claim that free particles define paths that are straight in the sense of the Desargues configuration. The coordinates x^μ are special not just because the equation of motion expressed in terms of them takes the special simple form (5.1); the coordinates x^i $(i = 1, 2, 3)$ should also be special in relation to the metrical properties of space. When Newton talks of uniform speeds, he means equal distances being traversed in equal times, and these distances are meant in the sense of Euclid. The projective-geometric formulation of the first law of motion would be of limited interest if the projective 4-geometry was not 'compatible' with the Euclidean metric of 3-space. In other words, relative to the special coordinates x^i above, the metric tensor g_{ij} should take the form $g_{ij} = \text{diag}(1, 1, 1)$. Indeed, significant efforts have been made to elucidate the geometric structures that underlie Newtonian mechanics in all its richness[17]. But note that the standard account in the literature posits *ab initio* a privileged foliation of the space-time manifold that rests on the existence of absolute simultaneity in the theory. There is a sense in which such an account oversteps the mark, and I want to dwell on this point momentarily.

2.2.3 Newtonian Time

'Absolute, true, and mathematical time, of itself, and from its own nature, flows equably without relation to anything external, and by another name is called duration . . .'[18]

What is meant when it is said that Newtonian time is 'absolute'? Many things.

One sense is that time flows independently of the existence of matter—if the material universe ceased to exist there would still be time. This is a contentious issue that is not our concern now. A weaker notion is that time is in some sense intrinsically tied up with *change* in the arrangement of matter, that it has a metric character, and that this metric character does not depend on the contingencies of the occasion. According to this notion of *duration*, it is the same everywhere and at all epochs of the universe.

[17] See e.g., Havas (1964), Anderson (1967), Earman and Friedman (1973), and Anderson (1990). It is interesting that in most (all?) of these accounts, the principle of inertia on its own is associated with the existence of an affine, rather than projective structure of space-time. I return to this point below, but note here that the extra structure being appealed to is the existence of 'affine' parameters τ in (2.1) such that the RHS of (2.2) vanishes in arbitrary coordinate systems and arbitrary reparametrization.

[18] Newton (1999).

It is a notion dealt with by Newton with remarkable sophistication, more, arguably, than was the case with Einstein when in 1905 he also assumed the existence of a temporal duration read by stationary ideal clocks. Newton did not identify the temporal metric with the behaviour of any given clock—indeed, he was aware (perhaps more than many of his followers!) of the limitations of real clocks. In the highly denuded world that we have been discussing of empty space populated solely by free particles, the only clock available is the free particle itself—one such particle is arbitrarily chosen, and the temporal intervals during which it traverses equal distances are taken to be equal.[19] Newton was fully aware that in the real world, no such inertial clocks exist, and indeed that *any* clocks that are subsystems of the universe may fail to march precisely in step with absolute time. 'It may be, that there is no such thing as an equable motion, whereby time might be accurately measured. All motions may be accelerated and retarded, but the flowing of absolute time is not liable to any change.'

What is the significance of this notion of time? At the end of the nineteenth century, a number of commentators, such as Auguste Calinon and Henri Poincaré in France and independently G. F. FitzGerald in Ireland, articulated the key notion. Physical time has to do with the choice of a temporal parameter relative to which the fundamental equations of motion of the isolated system under investigation take their simplest form.[20] Such a dynamical notion was already appreciated by Newton, and it had to do with his explicit association of absolute time in the *De Motu* with 'that whose equation astronomers investigate'. In practice, astronomers would use the rotation of the earth until the late nineteenth century as a clock, but Newton already foresaw the fact that absolute time cannot be *defined* in terms of the sidereal day.[21] He anticipated the notion of 'ephemeris' time which would be employed by the astronomers prior to the advent of atomic clocks. This issue will be revisited, along with the question of what a clock actually is, later.

Another absolute aspect of Newtonian time is that in so far as it is read by clocks, this reading is achieved without regard to the inertial state of motion of the clock. Better put, it is that proper time and coordinate time coincide: there is no time dilation. Special relativity rejects this feature of Newtonian time, as well as another. The last feature is ***absolute distant simultaneity***, which itself has two facets: (i) the the absolute nature of simultaneity relative to the frame in which the laws are postulated, and (ii) the invariance of this simultaneity relation under boosts. In relation to facet (i), we encounter the subtle business that has been such a prominent feature in the discussion of inertial coordinate systems in special relativity: *how to spread time through space*. The conventional nature of distant simultaneity in special relativity—not to be confused with the relativity

[19] This point was first given its due prominence in the work of Neumann (1870), and later Lange (1886). (It was Lange who coined the term 'inertial system'.) For discussion of their contributions see Barbour (1989), pp. 654–7.

[20] See Calinon (1897), Poincaré (1898), and FitzGerald (1902).

[21] See Barbour (1989), p. 633.

of simultaneity—is one of the most hotly debated issues in the literature. But the issue is not the exclusive remit of relativity theory, and the common claim that Einstein revolutionized the notion of time seems to me to be overstated, at least in regard to facet (i) of Newtonian time. Einstein's contribution will of course be discussed below, and my scepticism will be spelt out in detail. What I want to underline here is that even with the introduction of 3-dimensional Euclidean metric structure into the Newtonian 4-manifold, the simple equations for the free particle

$$d^2x^i/dt^2 = 0 \tag{2.5}$$

($i = 1, 2, 3$) are form invariant, or *covariant*, under the linear transformations of the form

$$\mathbf{x}' = \mathbf{x}; \quad t' = t - \vec{\kappa}.\mathbf{x}, \tag{2.6}$$

for any constant 3-vector field $\vec{\kappa}$. These transformations are not elements of the 10-parameter Galilean group (even of the subgroups that don't involve boosts); they represent a change in the assignment of the simultaneity relation between events. (Note that the locus of events for which $t' = 0$ corresponds to that for $t = \vec{\kappa}.\mathbf{x}$.) In the rarified Newtonian world of *free* particles moving in Euclidean 3-space, there is no privileged notion of simultaneity, even when viewed with respect to the frame at rest relative to Newton's hypothetical absolute space.

It follows that Newtonian simultaneity is a by-product of the introduction of forces into the theory. Indeed, Newton spread time through space in inertial frames in such a way that actions-at-a-distance like gravity are instantaneous and do not travel backwards in time in some directions. It is a highly natural convention—it would be barmy to choose any other—but it is a convention nonetheless, and one consistent with the standard Galilean coordinate transformations. The only appearance of something like this claim in the literature that I am aware of is due to Henri Poincaré.[22] In his remarkable 1898 essay *The Measure of Time*, Poincaré asserts that statements involving the order of occurrence of distant events have no intrinsic meaning—their meaning only being assigned by convention. Note that part of his argument relates to what we now call the Cauchy, or initial data problem in Newtonian mechanics. Poincaré imagines a toy system of three bodies: the sun, Jupiter, and Saturn represented as mass points. He remarks that it would be possible to predict future (and past!) behaviour of the system by taking the appropriate data concerning Jupiter (and presumably the sun) at the instant t, together with data concerning Saturn not at t but at $t + a$. This odd procedure would involve using 'laws as precise as that of Newton, although more complicated'. Poincaré asks why the unorthodox 'aggregate' of positions

[22] For a fascinating account of the extraordinary range of Poincaré's accomplishments—from mining engineering to the highest reaches of abstract mathematics—see Galison (2004).

and velocities is not regarded as the cause of future and past aggregates, 'which would lead to considering as simultaneous the instant t of Jupiter and the instant $t + a$ of Saturn'. In answer there can only be reasons, very strong, it is true, of convenience and simplicity.[23]

Related brief remarks appear in the chapter on classical mechanics in his 1902 *La Science et l'hypothèse*:

There is no absolute time. When we say that two periods are equal, the statement has no meaning, and can only acquire a meaning by convention. ... Not only have we no direct intuition of the equality of two periods, but we have not even direct intuition of the simultaneity of two events which occur in two different places.[24]

In 1898, Poincaré had already written 'The simultaneity of two events, or the order of their succession, the equality of two durations, are to be so defined that the enunciation of the natural laws may be as simple as possible.'[25]

I will return to this delicate issue in Chapter 6, where a more systematic discussion is given of the conventionality of distant simultaneity in both Newtonian and relativistic kinematics. In conclusion, I want to emphasize a few points in relation to Newtonian time.

First, the fact that under the Galilean transformations the notion of Newtonian simultaneity is frame-independent, so that whether two events occur at the same time does not depend on the state of motion of the observer, does not mean that a conventional element related to spreading time through space is absent in

[23] Poincaré (1898), section XI. Galison (2004) urges the importance of understanding Poincaré's 1898 essay within the context of his active role as a member of the Paris Bureaux des Longitudes. Galison correctly stresses that Poincaré's context is very different from Einstein's (p. 44). In fact Poincaré was immersed in 'the concerns of real-world engineers, sea-going ships captains, imperious railway magnates, and calculation-intensive astronomers' (p. 165). But Galison's absorbing analysis is marred by two features. (i) In his lengthy discussion of the new telegraphic method of establishing longitude, and Poincaré's intimate knowledge of it, Galison repeatedly seems to associate the conventionality of the technique with the fact that the time of transmission of the signals had to be taken into consideration in establishing simultaneity between distant sites (see particularly pp. 182–3, 189–90). In this sense, the electric mappers 'did not need to wait for relativity'. But the basis for the conventionality of distant simultaneity that Einstein espoused was not simply the fact that light signals used to synchronize distant clocks take time to propagate; what is at issue is *how* the transit time is dealt with. Galison gives a nice account of the simple method electric surveyors used to 'measure' the time it took a telegraphic signal to pass through cables (p. 184). Where then is the relevant conventional element? It happens to be in the crucial assumption that the velocity of the electric signal is the same in both directions. Now Galison is aware that this assumption *inter alia* lies behind the standard measurement of transit time (p. 186), but its role in his analysis seems to be of a secondary nature, almost an afterthought. In short, it is not easy to reconcile Galison's reconstruction of Poincaré's conventionalism about simultaneity with Poincaré's discussion in the 1898 essay of both telegraphy and the initial value problem in astronomy. (ii) Poincaré was a conventionalist, as we have seen, about both simultaneity (spreading time through space) and the temporal metric used in physics. In both cases, the choice depends on convenience. But the two issues are entirely distinct; the former involves synchronizing distant ideal clocks, whereas the latter concerns the very meaning of a single ideal clock. At times Galison seems to treat the two issues as one (see particularly pp. 187, 239).

[24] Poincaré (1952), chap. VI.

[25] Poincaré (1898), §XIII.

Newtonian mechanics. It is a feature of any theory of motion. Second, the convention standardly adopted in Newtonian mechanics is motivated by the structure of forces in the theory—it is completely obscure in relation to inertial effects alone.

2.2.4 Newtonian Space

Euclidean 3-space was introduced rather blithely above, despite the fact that it is not required, or so it is often claimed, in the attempt to explicate the conspiracy that is inertia. The point I made earlier was that the conspiracy couched in the language of projective or affine spaces is highly significant, but it is not the real McCoy. The full significance of the principle of inertia, and of the inertial frame, incorporates the notion of 3-dimensional spatial distance.

> It is on the possibility of measuring distance that ultimately the whole of dynamics rests. All the higher concepts of dynamics—velocity, acceleration, mass, charge, etc.—are built up from the possibility of measuring distance and observing motion of bodies. Examination of the writings of Descartes and Newton reveals no awareness of the potential problems of an uncritical acceptance of the concept of distance. Both men clearly saw extension as something existing in its own right with properties that simply could not be otherwise than as they, following Euclid, conceived them. As Newton said in *De gravitatione*: ... 'We have an exceptionally clear idea of extension.'[26]

But even before one considers the threat to this clear idea caused by the emergence of non-Euclidean geometry, the more one thinks about the physical notion of distance, the more elusive it becomes. (Einstein had a lifelong struggle with the notion of a metric 3-space, his confidence in defining it in terms of the physics of packing mobile rigid bodies waxing and waning.)[27] In a counterfactual Newtonian world comprising just a collection of N mass points interacting gravitationally, it is possible in principle to describe the motion of the particles using the Newtonian laws of motion. We can imagine such a world because in ours we have access to rigid, or near-rigid bodies that allow us to give those all-important numbers associated with the separation between points a more or less direct operational meaning. But in the hypothetical world of unaggregated point particles, the deep structural feature of the time-evolving configuration of the totality of particles that is Euclidean space has no meaning except in and through that dynamical evolution. But meaning it has, and it may be worth bearing this point in mind when assessing the claim Poincaré made in 1902: 'If... there were no solid bodies in nature there would be no geometry.'[28]

It will be assumed, at any rate, in the following that rigid rulers that come to rest with respect to the inertial frame, and hence have no significant net external forces acting them, are objects which also come rapidly to internal

[26] Barbour (1989), p. 692.
[27] See Brown and Pooley (2001), Ryckman (2005), §3.3, and Paty (1992).
[28] Poincaré (1952), p. 61.

equilibrium and stay there. Furthermore, they are capable to a 'high' degree of accuracy of reading off spatial intervals defined by the Euclidean metric. Note that I do not assume that the metric is *defined* by the behaviour of rigid bodies.

2.2.5 The Role of Space-time Geometry

In their influential 1973 article on Newton's first law of motion, John Earman and Michael Friedman claimed that no rigorous formulation of the law is possible except in the language of 4-dimensional geometric objects. But the appearance of systematic studies of the 4-dimensional geometry of Newtonian space-time is relatively recent; the first I am aware of is a 1909 paper by P. Frank immediately following the work of Minkowski on the geometrization of special relativity.[29] It is curious that so much success had been achieved by the astronomers in applying Newton's theory of universal gravity to the solar system (including recognizing its anomalous prediction for the perihelion of Mercury) well before this date. How could this be if the astronomers were unable to fully articulate the first law of motion, and hence the meaning of inertial frames?

How tempting it is in physics to think that precise abstract definitions are if not the whole story, then at least the royal road to enlightenment. Yet consider the practical problem faced by astronomers in attempting to fix the true motions of the celestial bodies. The astronomers who know their Newton are not helped by the further knowledge that Newtonian space-time comes equipped with an absolute flat affine connection. Even Newton realized that his absolute space and time—those entities distinct from material bodies but whose existence is necessary in Newton's eyes to *situate* the bodies so that their motions can be defined—'by no means come under the observation of our senses'. Newton was keenly aware of the need to arrest the backsliding into pure metaphysics: but how to make the theoretical edifice touch solid empirical ground? The story as to how this was achieved in practice, involving the contributions of such men as Neumann, Lange, and Tait, is both fascinating and far too little known.[30] But it happened with little more than the knowledge of the nature of the gravitational force and the elements of Euclidean spatial geometry, and growing amounts of astronomical records.

Let us pass on to a more philosophical question regarding the role of geometry in our understanding of motion. In Newtonian space-time, the world-lines of free particles are, as we have seen, widely regarded as geodesics (straights) of the postulated affine connection; in Minkowski space-time they are geodesics of the Minkowski metric. What is geometry doing here—codifying the behaviour of

[29] Frank's work is cited in Havas (1964), fn. 13.

[30] An eloquent introduction to this story is found in Barbour (1989), chap. 12, which deals with the empirical definitions of inertial system by Lange, of the equality of time interval by Neumann and that of mass by Mach. Barbour (1999) discusses the contribution of Tait.

free bodies in elegant mathematical language or actually explaining it? It is widely known that Einstein's first reaction to his ex-professor Hermann Minkowski's geometrization of his special theory was negative. But as Einstein's ideas on gravity developed and the need for changing its status from a Newtonian force to something like geodesic deviation (curvature) in four dimensions became clear to him, his attitude underwent a fundamental shift. As late as 1924, in referring to the special theory and its treatment of inertia, he wrote 'That something real has to be conceived as the cause for the preference of an inertial system over a noninertial system is a fact that physicists have only come to understand in recent years.'[31]

Einstein's position would, three years later, undergo yet another shift, at least in relation to inertia in the general theory, but that is a story for later in the book. The idea that the space-time connection plays this explanatory role in the special theory, that affine geodesics form ruts or grooves in space-time which somehow guide the free particles along their way, has become very popular, at least in the late twentieth century philosophical literature. It was expressed succinctly by Nerlich in 1976:

> [W]ithout the affine structure there is nothing to determine how the [free] particle trajectory should lie. It has no antennae to tell it where other objects are, even if there were other objects ... It *is because space-time has a certain shape that world lines lie as they do*.[32]

It is one of the aims of this book to rebut this and related ideas about the role of absolute geometry. Of course, Nerlich is half right: there is a prima facie mystery as to why objects with no antennae should move in an orchestrated fashion. That is precisely the pre-established harmony, or miracle, that was highlighted above. But it is a spurious notion of explanation that is being offered here. If free particles have no antennae, then they have no space-time feelers either. How are we to understand the coupling between the particles and the postulated geometrical space-time structure? As emerges from a later discussion of the geodesic principle in general relativity, it cannot simply be in the nature of free test particles to 'read' the projective geometry, or affine connection or metric, since in the general theory their world-lines follow geodesics *approximately*, and then *for quite different reasons*.

At the heart of the whole business is the question whether the space-time explanation of inertia is not an exercise in redundancy. In what sense then is the postulation of the absolute space-time structure doing more explanatory work than Molière's famous dormative virtue in opium? It is non-trivial of course that inertia can be given a *geometrical* description, and this is associated with the fact that the behaviour of force-free bodies does not depend on their constitution: it is universal. But what is at issue is the arrow of explanation. The notion of explanation that Nerlich offers is like introducing two cogs into a machine which

[31] Einstein (1924). [32] Nerlich (1976), p. 264.

only engage with each other.[33] It is simply more natural and economical—better philosophy, in short—to consider absolute space-time structure as a codification of certain key aspects of the behaviour of particles (and/or fields). The point has been expressed by Robert DiSalle as follows:

> When we say that a free particle follows, while a particle experiencing a force deviates from, a geodesic of spacetime, we are not explaining the cause of the difference between two states or explaining 'relative to what' such a difference holds. Instead, we are giving the physical definition of a spacetime geodesic. To say that spacetime has the affine structure thus defined is not to postulate some hidden entity to explain the appearances, but rather to say that empirical facts support a system of physical laws that incorporates such a definition.[34]

This theme will reappear later in the discussion of Minkowski space-time in special relativity.

2.2.6 Quantum Probes

We finish this section with a word about space-time structure seen from the perspective of quantum, rather than classical probes, or test bodies. The claim that the behaviour of free bodies does not depend on their mass and internal composition has been referred to as the 'zeroth law of mechanics.'[35] Consider also the weak equivalence principle (WEP) which can be stated like this. 'The behaviour of test bodies in a gravitational field does not depend on their mass and internal composition.'[36]

Now it is interesting that non-relativistic quantum mechanics violates both principles.[37] In the case of the zeroth law, this is seen merely by noting that the time-dependent Schrödinger equation for a free particle represented by the wavefunction $\psi = \psi(\mathbf{x}, t)$

$$i\hbar\frac{\partial\psi}{\partial t} = -\frac{\hbar^2}{2m}\nabla^2\psi \tag{2.7}$$

contains the mass m of the system. A more striking way of making the point is by way of the spread of the free wavepacket in empty space. If it is assumed that the packet is originally Gaussian with standard deviation (width) a, then after time t the width becomes

$$\Delta x = a\left(1 + \frac{\hbar^2 t^2}{4m^2 a^4}\right)^{1/2}. \tag{2.8}$$

[33] I am grateful to Chris Timpson for this simile.

[34] DiSalle (1995); see also DiSalle (1994).

[35] See Sonego (1995).

[36] See Will (2001), §2.1.

[37] See Sonego (1995). Sonego points out that the zeroth law is a special case of the WEP, and hence that a violation of the first is automatically a violation of the second.

How much the wavepacket spreads depends inversely on the (square of the) mass of the system.

Violation of the WEP was all but established experimentally in the 1970s with intriguing experiments using neutron interferometry, in which the two coherent beams inside the interferometer are made to experience different gravitational potentials (associated with a difference of height of the order of a centimetre!) as a result of the interferometer being tilted away from its usual horizontal orientation. The tilting produced a detectable loss of interference between the two beams, in good agreement with the predictions of quantum mechanics.[38] The effect is indeed predicted to depend on the (gravitational) mass of the particle.[39]

It is important to emphasize that despite all this, nothing in quantum theory threatens the claim that underlies the whole edifice of the general theory of relativity, namely that it is impossible for an observer to distinguish immersion in a homogeneous gravitational field from acceleration. (We return to this principle in Chapter 9.) Nonetheless, the violation in quantum theory of the zeroth law has led to doubts being expressed as to the truly geometric nature of the metric field in general relativity, as well as that of the affine connection in Newtonian space-time.[40] As far as the affine connection is concerned in both Newtonian and Minkowski space-time, it seems to me that use of quantal, as opposed to classical, test bodies merely makes this structure, not less geometrical *per se*, but simply of less direct operational significance. The real issue is not whether physical geometry is easy to get your hands on, but rather whether, when it is absolute and immune to perturbation as in Newtonian and Minkowski space-time, it offers a causal explanation of anything.

2.3 THE LINEARITY OF INERTIAL COORDINATE TRANSFORMATIONS

So far we have been looking at the nature of the inertial coordinate system x^μ, the existence of which was claimed to be tantamount to Newton's first law of motion. We may think of the associated inertial 'reference frame' S as the equivalence class of coordinate systems related to x^μ by rigid translations and rotations of space, and translations in time. It is clear that such linear transformations preserve both the form of (2.5) and the spatial and temporal metrics. But now the questions arise: are there other inertial frames corresponding to different states of motion, and if so, why are the transformations between them systems linear? Two approaches to answering these questions will be considered, each depending on the existence of certain dynamical symmetry principles.

[38] For a review of these experiments, see Werner (1994) and Audretsch *et al.* (1992).

[39] For a recent discussion of the behaviour of a quantum particle in free fall and its connection with the equivalence principle, see Davies (2004*a*, *b*).

[40] Sonego (1995).

The first approach relies on the relativity principle. Let us suppose that there is a reference frame moving rigidly with uniform velocity relative to the system x^{μ}. According to the relativity principle (of which much will be said in the following chapter) free bodies must also move uniformly and rectilinearly relative to some system x'^{μ} adapted to S'. Since whether a body is free or otherwise is frame-independent, we are interested now in the question as to what constraint on the functions f^{μ} in

$$x'^{\mu} = f^{\mu}(x^0, x^1, x^2, x^3) \tag{2.9}$$

is imposed by the requirement that (2.5) take the same form in the primed system of coordinates: uniform velocities must be transformed by the f_i into uniform velocities. It can be shown that the transformations take the form of a 'rational fraction', or 'projective' transformation:

$$x'^{\mu} = \frac{a^{\mu}{}_{\nu} x^{\nu} + a^{\mu}}{b_{\nu} x^{\nu} + b}, \tag{2.10}$$

again using the Einstein summation convention. Note that the denominator does not depend on the chosen index μ. Now if transformations take finite coordinates always into finite coordinates (equivalently, if coordinates are defined everywhere), then it follows from (2.10) that the transformations are linear. This approach was advocated by Fock, following Weyl.[41]

The second approach follows Einstein, who in his 1905 paper, grounds the linearity of the coordinate transformations associated with boosts on the homogeneity of space and time. Einstein provided no justification for this connection; here is a way of spelling it out.

[41] The main result here is taken from Fock (1969), Appendix A. See also Torretti (1983), pp. 75–6. Recently, Jahn and Sreedhar (2001) have claimed that the most general transformations that leave the Newtonian equations of motion for the free particle form invariant are given by

$$x_i' = \frac{R_{ij} x_j + \alpha_i + v_i t}{\gamma t + \delta}, \; t' = \frac{\alpha t + \beta}{\gamma t + \delta}, \tag{2.11}$$

where $\alpha\delta - \beta\gamma = 1$ and $R^T R = 1$. These are far more restricted than (2.10). The reason is that Jahn and Sreedhar are actually determining the most general transformations that are variational symmetries of the non-relativistic action

$$S = \frac{m}{2} \int \left(\frac{\mathrm{d}x_i}{\mathrm{d}t} \right)^2 dt, \tag{2.12}$$

which are more constrained than those that preserve the form of the Euler–Lagrange equation of motion (2.5) resulting from this action under Hamilton's principle of least action. (It is already widely known that variational symmetries of actions exist that are not symmetries of the equations of motion; see for example Brown and Holland (2004). What the present example is reminding us is that even where variational symmetries are dynamical symmetries, we must be careful not to conflate the maximal symmetry groups associated with actions and their associated Euler–Lagrange equations.) Note that the transformations (2.6) are not of the type (2.11), despite being a symmetry of (2.5).

We rewrite equation 2.9 as follows:

$$x'^{\mu} = f^{\mu}(X) \tag{2.13}$$

where X represents the quadruple $x^0, \ldots, x^3$. Let us suppose that the coordinate transformations encode information regarding the behaviour of moving rods and clocks. (This highly non-trivial supposition should not be taken as obvious, and will be justified in the next section.) Then such behaviour should not depend on where the moving rods and clocks find themselves in space, or in time, if the homogeneity claim holds. Consider now the infinitesimal interval $(dx^0, \ldots, dx^3)$ for which

$$\mathrm{d}x'^{\mu} = \frac{\partial f^{\mu}}{\partial x^{\nu}}\mathrm{d}x^{\nu}. \tag{2.14}$$

Homogeneity implies that the coefficients $\partial f^{\mu}/\partial x^{\nu}$ must be independent of X, which means that the f^{μ} must be linear functions of the coordinates x^{μ}.[42]

Which of these approaches is more fundamental depends in part on whether the relativity principle has the status of a fundamental postulate or of something derivable from more basic principles. In Chapter 7 we will examine arguments for the latter position; if valid, they lend support to the priority of Einstein's approach to establishing linearity.

2.4 THE ROD AND CLOCK PROTOCOLS

Now consider a 'rigid' rod lying in a direction parallel with the x-axis in the frame S with rest length L_0, but now boosted to velocity v in that direction relative to S. We assume of course that following the application of the force that has been applied to the rod to boost it, the rod has time to retain an equilibrium configuration. According to *the rod protocol*, we define the longitudinal length change factor of the rod relative to S as

$$C_{\|} = \Delta x / L_0, \tag{2.15}$$

where Δx is the length of the moving rod relative to S, which is taken to mean the distance between events that occur simultaneously at the extremities of the rod. This distance can be given direct operational significance by way of stationary rulers in S.

[42] See, e.g., Terletskii (1968), pp. 18–19, and Lévy-Leblond (1976). A lengthier, and possibly more rigorous proof of this kind is found in Berzi and Gorini (1969).

Similarly we can define the transverse length change factor $\mathcal{C}_\perp$ of the same rod but this time lying perpendicular to its motion along the x-axis:

$$\mathcal{C}_\perp = \Delta y/L_0 = \Delta z/L_0. \tag{2.16}$$

Note that we are assuming here that the length change factor is isotropic in the transverse plane. In particular we are assuming the factor is the same whether the direction involved is given by the y-axis or the z-axis.

Now according to *the clock protocol*, we define the time dilation factor in the following manner. Consider an ideal clock also moving at speed v along the x-axis, and two events occurring on the clock's worldline, with τ being the (proper) time read off by the clock between the two events. Relative to S, the two events clearly occur at different spatial locations, and the (coordinate) time interval between them, denoted by Δt, is established by the difference in readings of stationary, synchronized clocks at those locations. Then the time dilation factor for the clock relative to S is given by

$$\mathcal{D} = \Delta t/\tau. \tag{2.17}$$

It is important to stress at this point that both protocols are defined with respect to the clock synchrony convention associated with the frame S; change the way clocks are synchronized and the factors will in general change. But the corresponding convention associated with frame S'—the rest frame of the moving rod or clock—is irrelevant. Indeed, *the protocols are entirely defined relative to a single frame.*[43] In the case of the length factors $\mathcal{C}_\parallel$ and $\mathcal{C}_\perp$, it is also important to realize that the overall distortion effect (if any) that they describe should not be conflated with the *visual* shape change induced by motion. The apparent shape of the moving body that an observer 'sees' is the result of light emitted from different points on the body's surface entering simultaneously into his or her eyes, or into a camera, and this generally will be quite different from the distortion associated with the rod protocol![44] In the case of the time dilation factor, it is important not to lose sight of the fact that we are comparing the proper time read off by *one* moving clock against the coordinate time defined by *two* synchronized stationary clocks.

[43] Thus the contraction and dilation factors do not depend on, nor are an automatic result of, the relativity of simultaneity, despite occasional claims to the contrary. (In Galison (2004), for instance, it is claimed (p. 22) that the relativity of simultaneity leads to a relativity of lengths.)

[44] The first discussion of this issue, given by A. Lampa in 1924, seems to have been widely overlooked; the problem was rediscovered in 1959, and then early contributors were J. Terrell, R. Penrose, R. Weinstein, V. Weisskopf, and M. Boas. It is now known that bodies subtending a small solid angle appear rotated when moving at relativistic speeds, and bodies subtending a large solid angle appear both rotated and distorted. A useful bibliography is found in Kraus (2000), where the separate Doppler and searchlight effects are compared with the geometric aberration effect.

Now we raise the simple question as to what the length change and time dilation factors have to do with the transformations between inertial coordinate systems. To this end, we introduce two further aspects of the protocols. We assume

1. *The universality of the behaviour of rods and clocks.* The length change factors for rods and the dilation factor for clocks do not depend on the constitution of these objects, nor on the means by which the rods and clocks were boosted.
2. *The boostability of rulers and clocks.* Any object that can act as a rigid ruler in the frame S when stationary relative to that frame retains that role in its new rest frame S' when boosted. The same assumption holds for ideal clocks.

Assumption 1 represents the second great miracle in our story after the existence of inertial frames, at least if motion has any affect at all on rods and clocks. It is highly nontrivial, indeed remarkable, that such behaviour should be insensitive to the microscopic make-up of the bodies in question. We are now so used to this miracle that it seems mundane, but it is worth recalling that in the early twentieth century the Michelson–Morley experiment was repeated on several occasions with different substances making up the rigid support of the interferometer mirrors precisely to test for the universality of the FitzGerald–Lorentz deformation effect. Later in this book we shall also see why it is non-trivial that length change and dilation factors should likewise be independent of the manner in which rods and clocks, respectively, are boosted.

Assumption 2 is more of a stipulation than an assumption. Our earth-bound laboratories are not in a constant state of inertial motion; their velocities relative to the sun for example are changing constantly. Yet we neither want nor expect to have to change the instruments in our laboratories every few days for fear that rigid standards of length etc. are corrupted simply because they are slowly changing their state of motion.

It is easy to infer from Assumptions 1 and 2 that the rest length of any rigid rod will be invariant under the linear coordinate transformations (2.7). Now the two events mentioned above occurring simultaneously at the end-points of the moving rod will in general not be simultaneous relative to S', but since the rod is at rest in that frame, the distance $\Delta x'$ between them will nonetheless coincide with the invariant rest length. So we can rewrite (2.15) and (2.16) as

$$\mathcal{C}_{\parallel} = \Delta x/\Delta x' \tag{2.18}$$

$$\mathcal{C}_{\perp} = \Delta y/\Delta y' = \Delta z/\Delta z'. \tag{2.19}$$

Similarly we can rewrite (2.17) as

$$\mathcal{D} = \Delta t/\Delta t'. \tag{2.20}$$

It is easy then to show that the linear coordinate transformations between inertial frames S and S' with their adapted coordinate systems in the standard configuration can, given (2.18–2.20), be written in total generality in the following form:

$$x' = \frac{1}{\mathcal{C}_{\parallel}}(x - vt) \tag{2.21}$$

$$y' = \frac{1}{\mathcal{C}_{\perp}}y \tag{2.22}$$

$$z' = \frac{1}{\mathcal{C}_{\perp}}z \tag{2.23}$$

$$t' = \frac{1}{(1 - \alpha v)\mathcal{D}}(t - \alpha x). \tag{2.24}$$

The factors $\mathcal{C}_{\parallel}$, $\mathcal{C}_{\perp}$, and $\mathcal{D}$ will in general depend on the velocity v of the frame S' relative to S, and we expect that that as $v \longrightarrow 0$, these factors will tend to 1. Now $t' = 0$ implies that $t = \alpha x$, so the factor α in (2.24) gives the slope of the simultaneity planes defined in S' as they appear in S. Thus α can be called *the relativity of simultaneity factor*, and unlike the other factors it depends on how distant clocks are synchronized in *both* frames S and S'. Furthermore $\alpha \longrightarrow 0$ as $v \longrightarrow 1$.

We note that the velocity transformation rules that follow strictly from (2.21–2.24) are these:

$$u'_{x'} = \frac{(1 - \alpha v)(u_x - v)\mathcal{D}}{(1 - \alpha u_x)\mathcal{C}_{\parallel}} \tag{2.25}$$

$$u'_{y'(z')} = \frac{(1 - \alpha v)u_{y(z)}\mathcal{D}}{(1 - \alpha u_x)\mathcal{C}_{\perp}} \tag{2.26}$$

where u_x, u_y, and u_z are the components of velocity of some body or signal as measured relative S, and $u'_{x'}$, $u'_{y'}$, and $u'_{z'}$ are the components of velocity of the same system as measured relative S'. If now we denote by V' the velocity of frame S relative to S', then $V' = u'_{x'}$ when $u_x = 0$. So from (2.25) we obtain

$$V' = -(1 - \alpha v)v\mathcal{D}/\mathcal{C}_{\parallel}. \tag{2.27}$$

Thus, the condition of *Reciprocity* holds (which means that $V' = -v$) if and only if

$$\mathcal{C}_{\parallel} = (1 - \alpha v)\mathcal{D}. \tag{2.28}$$

One sees that Reciprocity is sensitive to the synchrony conventions adopted in *both* frames S and S'.[45]

Finally, for some body moving along the positive x-direction relative to S with acceleration $\dot{u}_x$, the acceleration transformation resulting from (2.25) is

$$\dot{u}'_{x'} = \frac{(1 - \alpha v)^3 \dot{u}_x \mathcal{D}^2}{(1 - \alpha u_x)^3 \mathcal{C}_{\parallel}} \tag{2.29}$$

where u_x is the instantaneous velocity of the body relative to S. The acceleration $\dot{u}'_{x'}$ is generally a function of t', even when the acceleration relative to S is uniform.

[45] In case the reader finds any deviation from Reciprocity to be inconceivable, it may be worth noting that in special relativity, the principle of reciprocity for the relative speed between an inertial observer and one being uniformly accelerated holds only at events where their world-lines coincide. See Hamilton (1978).

3

The Relativity Principle and the Fable of Albert Keinstein

All manner of fable is being attached to my personality, and there is no end to the number of ingeniously devised tales.

Albert Keinstein, with apologies to Albert Einstein.[1]

Albert Einstein was not the first to use the relativity principle (RP) as a postulate in the treatment of a problem in physics. Christian Huygens had done so over two hundred years earlier in his treatment of collisions. And as we shall see shortly, even Newton in his pre-*Principia* writings viewed the RP as having the same axiomatic status as his laws. But the prominent role played by the principle in Einstein's 1905 paper on moving bodies in electrodynamics marked the beginning of a new attitude concerning the foundational status of symmetries in physics.

For Einstein, the RP was to be interpreted, at least in his 1905 paper, as a phenomenological principle analogous to the well-known laws of thermodynamics. There was something new and something old in Einstein's RP. It was new (though due acknowledgement must be given to Henri Poincaré, who had actually anticipated Einstein in this respect) to the extent that it encompassed electromagnetic phenomena and hence optics. After all, what was nontrivial in 1905, given all that had happened in physics in the nineteenth century, was deciding not the validity of the principle but its scope. Yet it was old in the sense that all that Einstein was really doing in extending its scope to electrodynamics was to restore the principle to its original Newtonian form. It is that form that interests us now.

3.1 THE RELATIVITY PRINCIPLE: THE LEGACY OF GALILEO AND NEWTON

3.1.1 Galileo

In a celebrated 1632 *Dialogo* thought experiment,[2] Galileo invites us to imagine diverse experiments—involving leaking bottles, small flying animals, etc.—being

[1] Calaprice (1996), p. 13. [2] Galilei (1960).

performed in the cabin of a stationary ship. He tells us that no difference in the outcomes of the experiments will be discernible if they are repeated when the ship is sailing uniformly over calm waters in a straight line.

The process of putting the ship into motion corresponds (with an important caveat to be elucidated below) to what today we call an active pure boost of the laboratory. A key aspect of Galileo's principle that we wish to highlight is this. For Galileo, the boost is a clearly defined operation pertaining to a certain subsystem of the universe, namely the laboratory (the cabin and equipment contained in it). The principle compares the outcome of relevant processes inside the cabin under different states of inertial motion of the cabin relative to the shore. It is simply assumed by Galileo that the same initial conditions in the cabin can always be reproduced. What gives the relativity principle empirical content is the fact that the differing states of motion of the cabin are clearly distinguishable relative to the earth's rest frame.

Let us delve a little deeper into what that content is. It is now well known that Galileo's relativity principle, like his principle of inertia, is associated with a class of horizontal motions, i.e. a class of motions of the laboratory that hug the earth's surface.[3] However, it is clearly implicit in Galileo's thinking that the regularities in the processes he discusses *do not depend on the time of year*. For a defender of Copernicanism such as Galileo, it would seem to follow that a second type of relativity principle must hold. This is one defined for certain states of motion relative not to the earth but to the 'fixed' stars, such as those instantiated by the earth in the course of its annual trek around the sun. But it is at this point that the weakness in Galileo's theory of motion becomes apparent. It is far from clear how to reconcile such insensitivity of the phenomena occurring inside the cabin to the seasons with Galileo's ill-fated theory of the tides. Recall that in this theory, the tides are caused in part by the existence of the earth's motion around the sun. It is ironic that today we rule out such a theory precisely on the grounds of the 'Galilean relativity principle'.

It should be noted that for some experiments explicitly mentioned in the ship thought-experiment, such as observation of the motion of drops of water in free fall, satisfaction of the relativity principle follows directly from Galileo's principle of inertia. Indeed, this case is a variant of the famous discussions concerning the behaviour of a cannon ball dropped either from the mast of a moving ship onto the deck, or from a tower fixed on the rotating earth. In all these cases, Galileo could explain the (supposed) vertical drop of the projectile by appeal to the inertial principle, and the assumption that perpendicular components of motion are independent. One might wonder whether Galileo did not view his relativity principle in its generality as a consequence of his principle of inertia. It is unlikely, though we cannot be sure. What is clear is that the inertial principle does not account, for example, for the velocity-independence of the *rate of flow of drops* in

[3] See, e.g., Chalmers (1993).

the case of the leaking bottle experiment. In fact, a little thought, aided no doubt by considerable hindsight, indicates that that for almost all the processes considered by Galileo in his thought-experiment, satisfaction of the relativity principle is not a mere consquence of the principle of inertia: the processes concern not just objects in motion but the dynamical mechanisms that produce that motion. (Why should our ability to throw a ball or jump around, or the mechanism used by butterflies to flap their wings, be independent of the state of motion of the cabin?)

Finally and signficantly, it seems likely that Galileo viewed his relativity principle as *universal in scope*, valid for all processes capable of being observed on the surface of the earth, including electrical and magnetic phenomena.[4] There is certainly no hint of the existence of counter-examples to the ship experiment in the *Dialogo*.

3.1.2 Newton

Largely as a result of the influence of Cartesian cosmology, post-Galilean seventeenth-century versions of the RP replaced the horizontally moving frames with rectilinearly moving ones. Here is Newton's formulation, as stated in Corollary V of the laws of motion in the *Principia*: 'The motions of bodies included in a given space are the same among themselves, whether that space is at rest, or moves uniformly forwards in a right line without any circular motion.'

Both Huygens and Newton espoused versions of the RP in which horizontally moving frames play no essential role, and both used the principle at times in a way that presaged Einstein's 1905 reasoning, viz. as a fundamental postulate which has non-trivial implications for the nature of dynamical interactions.[5] But Huygens, who believed in the relativity of *all* motion, was wisely silent on the crucial issue as to the whereabouts of bodies which constitute frames relative to which the Cartesian version of the inertial principle could be stated, and the relativity principle defined.[6] Newton, of course, appealed to an invisible, etherial body—absolute space—to this end. And by the time he wrote the *Principia*, the relativity principle for Newton had its status as a postulate demoted to that of a consequence of the laws of motion—a move which, as we shall see in the next subsection, resulted in one of the few *non sequiturs* in his great treatise.

It is noteworthy that at the end of his derivation of Corollary V, Newton explicitly mentions the 'proof' given by the 'experiment of a ship; where all motions [of bodies in the cabin] happen after the same manner, whether the ship is at rest, or is carried uniformly forwards in a right line.'[7] But unlike in the case of Galileo's

[4] See Brown (1993), p. 232.
[5] Barbour (1989), pp. 464–7, 571.
[6] See Barbour (1989), p. 462.
[7] Newton (1962).

RP, the question arises in relation to Newton's Corollary V whether the 'bodies included in a given space' cannot also refer to the whole material universe, assuming it is enclosed in a finite space. Given the role of absolute space in Newton's theory of motion, it certainly makes sense in the Newtonian scheme to talk about different states of inertial motion of the centre of mass of the physical universe. It is by no means clear that Newton would have regarded the relativity principle as inapplicable in this, untestable, case.

In this sense, as well as in the definition of the inertial motions involved, Newton's RP may be said to differ somewhat from Galileo's. But there are two important senses in which they concur.

First, in neither case does the statement of the principle presuppose the precise nature of the coordinate transformations between inertial frames. This point is enough to dispel what is sometimes regarded as a seemingly paradoxical feature of the RP, which in 1983 Michael Friedman put thus:

> The Newtonian principle of relativity is expressed in our freedom to transform coordinate systems by a Galilean transformation, the special [Einstein] principle of relativity in our freedom to transform coordinate systems by a Lorentz transformation, and so on. But this way of putting the matter makes the traditional relativity principles appear trivial. Are we not always free to transform coordinate systems by any transformation whatsoever? Will not any theory, regardless of its physical content, be *generally* covariant? If so, what physical content do the traditional relativity principles express?[8]

Friedman went on to develop a technical formulation of the traditional relativity principles (largely inspired by the 1960's work of James Anderson) designed to circumvent this kind of meltdown. But what Friedman overlooked in his analysis is the simple fact that the RP was born without any commitment to the nature of coordinate transformations at all.

The second point is this. For Newton, there were contact forces, and actions-at-a-distance; electrical and magnetic forces—examples of the latter—were in principle just as much part of his 'mechanics', and susceptible to his second law, as the gravitational force was. It is true that in the late seventeenth century what was missing was knowledge of the form of the electric and magnetic force expressions which would constitute the electromagnetic analogues of the 'inverse square' force expression for gravity. But nothing in the *Principia* suggests that the relativity principle is not applicable to electric and magnetic interactions.

There is, in short, probably no difference in scope between the relativity principles of Galileo and Newton, at least for terrestrial experiments, and no significant difference at all between those of Newton and the 1905 Einstein, at least for boosted *subsystems* of the universe.

[8] Friedman (1983), p. xii.

3.2 THE NON-SEQUITUR IN NEWTON'S COROLLARY V

Newton's proof of Corollary V, that is to say his derivation of the RP from the laws of motion, is cursory in the extreme. But it is clear that he was assuming two things.

The first was the Galilean transformations between inertial frames in relative motion, and hence the invariance of accelerations. (The Galilean transformations were simply taken for granted until the beginning of the twentieth century, not even being named until that period.) But significantly, Newton also presupposed the velocity independence of forces and masses. To remind ourselves as to why these two assumptions are needed we need to rehearse the meaning of Newton's second law of motion.

In its modern formulation, the law is $\mathbf{F} = m\ddot{\mathbf{x}}$, for a force acting on a body of inertial mass m.[9] Commentators who regard the equation as a *definition* of force underestimate its significance. For Newton's idea is that whatever the nature of the interaction involved, be it in virtue of contact with another body or action-at-a-distance, there should be a force expression—some vector-valued function analogous to the inverse square dependence on relative distance for gravity—which will slot into the place-holder **F** in the equation. The claim that there always exist such expressions is highly non-trivial (and, as finally realized in the twentieth century, false!). What Newton is assuming in Corollary V is first that for all the different kinds of interactions, the force expressions will have a property in common with the gravitational one. As measured by the observer at rest in the frame relative to which the laws of motion are initially postulated—let us call this the stationary frame—*the forces will not depend on the collective state of uniform motion of the system of bodies under consideration.* (Recall that the gravitational force depends on the relative distances between bodies, and not their absolute velocities.) A similar assumption is being made about the inertial and hence gravitational masses of the bodies.

Consider then an isolated system of bodies—a subsystem of the universe—enclosed in some region of space and interacting amongst themselves. Their accelerations relative to the stationary frame will be unaffected when the system of bodies as a whole is boosted. Relative to the observer comoving with the boosted system, the accelerations are the same as for the stationary observer, and likewise independent of the collective state of motion of the bodies. Indeed, given the same initial conditions, both observers will expect the same motions of the bodies to unfold 'among themselves'. This ensures the validity of the relativity principle and allows us to treat the forces and masses as invariant.

[9] It will be recalled that the acceleration $\ddot{\mathbf{x}}$ of the body is defined relative to the inertial frame arising out of the first law of motion. It is for this reason that the first law is not a special case of the second for $\mathrm{F} = 0$.

The trouble is that the velocity-independence of forces and inertial masses is not a consequence of the laws of motion, as Barbour noted in 1989.[10] Without this extra assumption, it is not possible to derive the relativity principle from Newton's laws and Galilean kinematics.[11]

3.3 KEINSTEIN'S 1705 DERIVATION

Let us fantasize that in 1705 a young German natural philosopher working in a patent office in Switzerland had come to the same conclusion after studying the *Principia*, and decided to invert the logic of Newton's argument.[12]

Albert Keinstein decided to *postulate* the RP, and investigate, like his illustrious near-namesake two hundred years later, how it constrains kinematics. As we shall see in Chapter 6, there are a number of routes he could have taken. The route he actually took in our story involved assuming explicitly the validity of Newton's laws of motion relative to the stationary (unprimed) frame, as well Newton's 'extra' condition concerning the velocity-independence of forces and masses relative to that frame. (Keinstein was aware that Newton had implicitly used the convention for synchronizing clocks in that frame according to which the gravitational interaction is instantaneous in every direction, or equivalently, the clock transport convention.)

It follows from the second law of motion and Newton's 'extra' assumption that relative to the stationary frame, the accelerations of a system of bodies enclosed in a given space are independent of the collective uniform rectilinear motion of the bodies (or of the motion of the space, as Newton would say). According to the RP, Keinstein surmised, the same velocity-independence also holds from the perspective of the moving (primed) frame. Indeed, given the same initial conditions in each frame, the same accelerations are predicted, and these are preserved under boosts of the systems of particles. It follows that accelerations are invariant across the two frames.

Keinstein realized that if you apply this conclusion to the general rule for transforming longitudinal accelerations given in equation (2.29), it follows (given the fact that v and u_x are freely chosen) that

$$\mathcal{D} = \sqrt{\mathcal{C}_{\|}}; \quad \alpha = 0. \tag{3.1}$$

[10] See Barbour (1989), pp. 31–2, 577–8, 608. It is somewhat hard to believe that recognition of this simple fact is not to be found in the literature prior to 1989, but I am unaware of any precursor.

[11] It might be thought that velocity-dependence of mass is a peculiarly relativistic phenomenon, but this view is becoming increasingly unpopular, as mass in special relativity is more and more identified with rest mass—which is of course invariant. This does not mean, however, that mass can be interpreted in special relativity as inertia; its meaning is now associated with the rest energy of a system. For reviews of the notion of mass in physics, see Okun (1989) and Roche (2005).

[12] The fable of Keinstein is taken, with minor modifications, from Brown (1993).

Thus a correlation between the values of the longitudinal length change factor and the time dilation factor has been established, but not the values themselves. Keinstein was at first tempted to argue that if Reciprocity is to hold, as one expects, then it follows from equation (2.28) and (3.1) that the mentioned factors can only be unity. Except for the transverse transformations (2.22) and (2.23), it would seem that the Galilean transformations have emerged. And regarding the transverse transformations, Keinstein could in principle have anticipated Einstein's (and independently Poincaré's) argument to the effect that the transverse length change factor is unity. This rather subtle argument will be discussed in Chapter 5; it depends only on the RP and spatial isotropy.

But Keinstein realized that appeal to the Reciprocity principle after deriving (3.1) is not innocent. After all, as we saw in section 2.4, once the distant clock synchrony convention is chosen in both the stationary and moving frame, Reciprocity becomes an empirical issue. Note that the second equation in (3.1) effectively fixes the synchrony convention in the moving frame, given Newton's implicit convention in the stationary frame. Is it really necessary to assume Reciprocity on top of the other postulates?

It was at this point that Keinstein realized that he had not made sufficient use of Newton's 'extra' assumption concerning the velocity-independence of forces and masses. What after all is a boosted rigid rod, if not a 'moving atomic configuration' to borrow the phrase Einstein used in 1949? The same phrase could have been used by Keinstein, in the spirit of Newtonian atomism. The atoms of the rod are held together rigidly by finding the stable equilibrium configuration consistent with the nature of the inter-atomic forces, whatever they are. That such is possible for the moving rod just as much as for the stationary one is a consequence of the relativity principle. But under the assumption that such unknown forces are themselves 'Newtonian', and therefore subject to the second law, it follows from Newton's 'extra' assumption that they, like the atomic masses, do not depend on the uniform state of motion of the rod relative to the stationary frame. Keinstein concluded that the shape and size of the rod is motion-independent: there can be no longitudinal nor transverse length change. Clearly, a similar conclusion holds in relation to clocks: there can be no time dilation.

This reasoning allowed Keinsten to put $\mathcal{C}_{\|} = \mathcal{C}_{\perp} = \mathcal{D} = 1$ in the transformations (2.21–2.24) to obtain

$$x' = x - vt \tag{3.2}$$

$$y' = y \tag{3.3}$$

$$z' = z \tag{3.4}$$

$$t' = \frac{t - \alpha x}{1 - \alpha v}. \tag{3.5}$$

Finally, Keinstein correctly reasoned that he could choose to synchronize clocks in the moving frame so as to preserve absolute simultaneity and render $\alpha = 0$, and thus obtain the Galilean transformations. Reciprocity did not after all have to be introduced as an extra assumption.

3.4 THE DYNAMICS–KINEMATIC CONNECTION

The Keinstein fable is of course preposterously anachronistic. It assumes that Keinstein understood the nature and meaning of kinematics better than anyone prior to Einstein. In the early eighteenth century it would have been wholly unclear to natural philosophers why any one would want to derive something (much later called the Galilean transformations) which was so self-evidently true. It assumes in particular that Keinstein had an understanding of the role of the clock synchrony convention in defining distant simultaneity that proved to be the last and hardest lesson Einstein had to learn to arrive at his 1905 derivation of the Lorentz transformations and which is an issue that still confuses people today.[13]

But the point of the fable is first to demonstrate that in pre-relativistic mechanics a derivation of the Galilean transformations is possible that mimics Einstein's 1905 derivation of the Lorentz transformations. Second, Keinstein's thinking demonstrates fairly clearly, I hope, that kinematics and dynamics are not independent departments of physics. As we will see later, the young Wolfgang Pauli stressed in 1921 that length contraction for instance is a 'very complicated process', and whether it takes place or not depends on the nature of the forces holding the constituent parts of the rod together. Keinstein was able to see that the absence of length contraction and time dilation is not a strict consequence of Newton's laws of motion, but is intimately connected with the 'extra' force and mass assumption that Newton employed in deriving the RP in Corollary V of the *Principia*.

[13] Lest it be thought that the conventionality of distant simultaneity is only an issue in relativity theory, recall the discussion in section 2.2.3 above.

4

The Trailblazers

> If the Michelson–Morley experiment had not put us in the worst predicament, no one would have perceived the relativity theory as (half) salvation.
>
> Albert Einstein[1]

> [M]y approach was not so terribly unsatisfactory. Lacking a general theory, one can derive some pleasure from the explanation of an isolated fact, as long as the explanation is not artificial. And the interpretation given by me and FitzGerald was not artificial. It was more so that it was the only possible one, and I added the comment that one arrives at the [deformation] hypothesis if one extends to other forces what one could already say about the influence of a translation on electrostatic forces. Had I emphasized this more, the hypothesis would have created less of an impression of being invented *ad hoc*.
>
> Letter of H. A. Lorentz to A. Einstein, 1915[2]

The cradle of the special theory of relativity was the combination of Maxwellian electromagnetism and the electron theory of Lorentz (and to a lesser extent of Larmor) based on Fresnel's notion of the stationary ether. Even when the search for detailed mechanical models of the electromagnetic ether had been given up as futile, the success of the program was enormous, and easy to underestimate today. In this chapter we look at the contributions of the principal figures concerned with the explanation of the *prima facie* surprising failure to detect any significant trace of the ether wind on the surface of the earth. It is well known that Einstein's special relativity was partially motivated by this failure, but in order to understand the originality of Einstein's 1905 work it is incumbent on us to review the work of the trailblazers, and in particular Michelson, FitzGerald, Lorentz, Larmor, and Poincaré.[3] After all they were jointly responsible for the discovery of relativistic kinematics, in form if not in content, as well as a significant portion of relativistic dynamics as well.[4]

[1] Einstein (1995). [2] Lorentz (1915).

[3] All five of these luminaries were born in the 1850s—what a decade for physics! But their deaths spanned over 40 years.

[4] The treatment of Michelson, Lorentz, and FitzGerald in this chapter is based largely on the study of the Michelson–Morley experiment and its immediate aftermath in Brown (2001).

4.1 MICHELSON

If it had not been for the prodding of Lord Rayleigh and H. A. Lorentz, Albert A. Michelson (1852–1931) might never had performed his celebrated 1887 ether wind experiment with Edward R. Morley. Of Polish birth, Californian upbringing, and a product of the US Naval Academy, Michelson had already established a reputation in experimental optics through his measurements of the velocity of light,[5] and in 1881 he had published the report of an intriguing ether wind experiment performed in Potsdam during a study leave in Europe. Michelson had designed a novel type of interferometer, and with money donated by Alexander Graham Bell had had it constructed by a reputable Berlin firm of instrument makers. Its behaviour would be surprising and ultimately galling for the precocious young American.

The Michelson interferometer involves a beam of near-monochromatic light being split by a semi-silvered mirror into two perpendicular beams which are then reflected back to the beam splitter; the light which reaches a telescope (eyepiece) close to the beam splitter is a superposition of contributions from the two arms. The two mirrors are equidistant from the beam splitter; suitable adjustment of the mirrors produces straight interference fringes. If the laboratory has speed v relative to the luminiferous ether, a rotation of the interferometer should produce a shift in the fringes, the effect being second order in the ratio v/c, where c is the speed of light relative to the ether. The very fact that the interferometer could produce a stable fringe pattern surprised the Continental leaders in the field of experimental optics; the device was one of unprecedented, almost implausible sensitivity.[6] In 1881 Michelson found no significant fringe shift under rotation in his Potsdam experiment.[7]

It is well known that the significance of this experiment lay in the fact that the predicted fringe shift is a second-order effect. A number of ether wind experiments had already been performed to test for first-order effects, all with null results. That this had not created a crisis in the ether theory was essentially due to the fact that the experiments all involved the transmission of light through transparent media, and there were excellent grounds, both experimental and theoretical, for the validity of the originally surprising value suggested by Fresnel in 1818 for the refractive index of media moving through the ether.[8] The Fresnel 'drag coefficient', much later understood to be a first-order consequence of the relativistic rule for transforming velocities under boosts, led to null results in every case. James Clerk

[5] A useful account of Michelson's many achievements in physics is found in Bennett *et al.* (1973).

[6] See Staley (2002).

[7] Michelson (1881). Michelson found that in the relatively bustling Berlin, where he first set up the interferometer, he could not keep the fringes stable during rotation. A useful discussion of the 1881 experiment can be found in Haubold *et al.* (1988).

[8] For a recent detailed study of the history of this development, see Pedersen (2000). In 1886 Michelson would perform, in collaboraton with Morley, a more accurate version of the famous 1851 Fizeau experiment confirming the value of the Fresnel coefficient.

Maxwell (1831–1879) had been the first to suggest the possibility of observing second-order ether wind effects, but doubted that they could be performed in terrestrial experiments. Michelson had risen to the challenge in 1881, or nearly.

Soon after he had published his 1881 paper, it was brought to Michelson's attention by the French physicist Alfred Potier that he had made an elementary error in calculating the passage time of light in the arm of the interferometer perpendicular to the motion of the laboratory through the ether. H. A. Lorentz would independently spot the error. Michelson had simply overlooked the effect of motion in the perpendicular arm, using the expression for the passage time that would hold were the interferometer at rest relative to the ether. The trouble was that when he corrected for the error, Michelson realized that the precision of the apparatus was just under that required to detect the fringe-shift. The 1881 interferometer was a technological marvel but it was a flop as an ether wind detector.

Michelson would complain that few people showed any interest in his 1881 experiment, and used this as an excuse for never publishing a note clarifying the miscalculation.[9] It is hard not to believe he was embarrassed by the episode, and wanted to move on. But some key figures rightly thought that the second-order effect was now too close to experimental test for the issue to be left to rest. Lord Rayleigh and Lorentz both encouraged Michelson to repeat the experiment using an improved (i.e. bigger) interferometer. The rest, as they say, is history. What is perhaps little known is the fact that the new expression for the transverse passage time in the celebrated 1887 paper with Morley is still wrong, but this time innocuously so: to second order it agrees with the correct one!

4.1.1 The Michelson–Morley Experiment Revisited

It is crucial in discussing the Michelson–Morley (henceforth MM) experiment[10] that the analysis take place from the perspective of some specific inertial frame S relative to which both the earth-bound laboratory is to a good approximation moving inertially, and the following assumption holds good.

The Light Principle (LP). *Relative to S, the two-way light speed in vacuo is a constant c, that is to say independent of the speed of the source and isotropic.*

Note that LP refers to the 'two-way' (or 'round-trip' or 'back-and-forth') light speed, whose determination can be established by a single clock, given knowledge of the distance the light traverses. At the end of the nineteenth century, this principle was widely accepted, under the assumption that S is the rest frame of the luminiferous ether. After all, if electromagnetic waves are propagating disturbances in this ether, their speed depends only on the elastic properties of the

[9] For a discussion of Michelson's attitude to his own ether-wind experiments, including that with Morley in 1887, see Staley (2002, 2003).

[10] Michelson and Morley (1887).

medium—and the ether is assumed to be homogeneous and isotropic. (The most prominent dissenter was Walter Ritz, who in the early years of the twentieth century would defend a ballistic theory of light in which the light speed does depend on the speed of the source.) Pauli would refer to the LP in 1921 as the 'true essence of the old aether point of view'[11] and as we see in the next chapter, Einstein in 1905 simply helped himself to it while performing the Wittgensteinian act of kicking away the justificational ladder of the ether. By the end of the century, Lorentz had already given up trying to give a mechanical underpinning to the ether, and just treated it as a new and somewhat mysterious form of imponderable matter. Nonetheless, relative to S the Maxwell–Lorentz equations of electromagnetism hold and no matter how etherial the ether proved to be, these equations support the LP.

The problem now is to calculate the passage times of light to-and-fro inside the two arms of the interferometer moving at speed v relative to S, say in the direction of the positive x-axis. Let us suppose arm A is pointing in this direction initially and arm B is pointing in the direction of the y-axis. Then it is easy to show given LP that the two-way passage times in the arms are relative to S:[12]

$$T_A = 2\gamma^2 L_A/c \tag{4.1}$$

$$T_B = 2\gamma L_B/c, \tag{4.2}$$

where L_A is the length relative to S of the moving arm A etc., and γ is the now familiar 'Lorentz' factor $(1 - v^2/c^2)^{-1/2}$.

Note that already something remarkable has taken place. Each arm of the MM interferometer, when at rest relative to S, is a Langevin clock, or would be if there were mirrors at both ends of the arm. Note that the passage times in (4.1) and (4.2) are not defined as the 'proper' periods of the 'longitudinal' and 'transverse' clocks, but correspond rather to the two values of the term Δt appearing in the expression (2.17) for the dilation factor of a clock. Supposing the Galilean transformations to hold, there is of course no dilation in either case. But this means that (4.1) and (4.2) coincide after all with the proper periods of the moving clocks, which are clearly not the same as the common rest periods of the clocks relative to S, viz. $2L_A/c$ and $2L_B/c$ respectively. So if the Galilean transformations are valid, the Langevin clock is therefore a counter-example to the boostability assumption for clocks made in section 2.4 above: its rest period is not invariant. This should sound warning bells about the Galilean transformations already.[13]

[11] Pauli (1981).

[12] Michelson and Morley in their 1887 paper give the value of T_B as $(2L_B/c)(1 + v^2/c^2)^{1/2}$; the mistake was first noted in the literature by Soni (1989). For speculation as to why this expression, which coincides to second order with (4.2) above, was arrived at see Capria and Pambianco (1994), fn. 6.

[13] This point is overlooked in attempts to provide a geometrical underpinning to the Maxwellian notion that electromagnetism is reconcilable with Galilean kinematics, as are found in Friedman (1983) and Earman (1989). It should be noted that these authors regard such attempts as doomed, but on other grounds.

But let us push on with the MM experiment. The rotation of the stone block supporting the apparatus allows one to compare the delay time $T_A - T_B$ with the corresponding delay time after rotation—a difference will show up as a shift in the interference fringes. On the assumptions that (i) the lengths of the arms of the interferometer were equal and unaffected by the motion relative to the ether, and (ii) the velocity of the laboratory relative to the ether is of the order of the orbital velocity of the earth relative to the sun, Michelson and Morley predicted a fringe shift of roughly 0.4 fringe for a rotation by $90°$. The actual result was 'certainly less than the twentieth part of this, and probably less than the fortieth part'.

The experiment has been repeated, in different guises and by different experimentalists, many times since 1887. Some of the most recent cases are mentioned at the end of the following chapter. But the rigorous theoretical analysis of the original experiment is more subtle than meets the eye. Two aspects are worth noting

(i) The standard derivation of the fringe-shift requires that the orientation of the semi-silvered mirror (beam splitter) is adjusted to deviate slightly from $45°$ in relation to the incoming beam of light relative to the stationary frame. It was only realized fairly recently that if this tiny deviation in tilt is not properly taken into account, both before and after rotation, then the exact formulas for the delay times are considerably more complicated than the standard ones, although the differences are only third and higher order in v/c.[14] And a separate 2002 analysis concluded that if, instead of the telescope in the experiment being focused on one of the mirrors, the interference pattern is formed on a vertical screen, then a first-order contribution to the fringe shift caused by rotating the interferometer would also be predicted on the basis of the usual assumptions.[15] In this case the observed absence of the fringe shift would not follow from either the FitzGerald–Lorentz deformation hypothesis (see below), or the Fresnel drag coefficient (since the effect is predicted *in vacuo*).

(ii) A common textbook explanation of the MM null result uses the Galilean rule for transforming velocities from the ether rest frame S to the rest frame S' of the laboratory, and then assumes that the rest length of the longitudinal arm undergoes contraction. This is incorrect, and very misleading—it is inconsistent with the relativity principle, and is based on an untenable picture of length contraction.[16] We shall see now that the MM result is directly inconsistent with the Galilean rule for transforming velocities.

[14] See Capria and Pambianco (1994) and Schumacher (1994); a discussion of the former paper in found in Brown (2001). Another related discussion of the reflection of light from moving mirrors is found in Gjurchinovksi (2004).

[15] See de Miranda Filho *et al.* (2002).

[16] Since according to the argument each arm of the interferometer changes its rest length under rotation, it follows that there is a violation of spatial isotropy relative to the lab frame S'. This is inconsistent with the relativity principle taken in conjunction with the fact that spatial isotropy is assumed to hold relative to the stationary frame S. Furthermore, the contraction is selective: it cannot hold for rigid rulers attached to the interferometer, since otherwise the contraction would be undetectable. A more complete discussion of this widespread, erroneous account of the null result is given in Brown (2001), section III.

4.2 MICHELSON–MORLEY KINEMATICS

What are the implications of the MM null result for the general inertial coordinate transformations (2.21–2.24)? The first and obvious implication is that *the two-way speed of light is isotropic relative to the lab frame* S',[17] just as it is assumed to be relative to the stationary frame S. This is a weaker claim than the invariance of the two-way speed! From the linearity of the coordinate transformations and the constancy of the light speed relative to S, it follows that the two-way light speed relative to S' is also independent of the speed of the source. So it is natural (but not compulsory) to adopt in *both* frames the Poincaré–Einstein convention for synchronizing clocks which renders the *one-way* speed of light isotropic. From this double convention alone, something interesting emerges.

Let us denote by c' the two-way speed of light relative to S'. Applying our synchrony convention in (2.25) and using $u_x = c \iff u'_{x'} = c'$, and $u_x = -c \leftrightarrow u'_{x'} = -c'$, one obtains

$$\alpha = v/c^2 \tag{4.3}$$

$$\mathcal{D} = c'\gamma^2\mathcal{C}_{\|}/c. \tag{4.4}$$

The relativity of simultaneity factor α is the familiar relativistic one, but so far no appeal has been made to the invariance of the light speed!

The second lesson of the MM null result concerns the length change factors. The delay times before and after rotation of the apparatus are given by

$$\Delta \equiv T_A - T_B = 2\gamma(\gamma L_A - L_B)/c \tag{4.5}$$

$$\Delta^{rot} \equiv T_A^{rot} - T_B^{rot} = 2\gamma(L_A^{rot} - \gamma L_B^{rot})/c, \tag{4.6}$$

where L_A^{rot} is the length relative to S of the rotated arm A, etc. Referring back to the discussion of the rod protocol in section 2.4 above, we see that the length change factors are given by

$$\mathcal{C}_{\|} = L_A/L_A^o = L_B^{rot}/L_B^o, \tag{4.7}$$

$$\mathcal{C}_{\perp} = L_B/L_B^o = L_A^{rot}/L_A^o, \tag{4.8}$$

where $L_{A(B)}^o$ is the length of arm $A(B)$ relative to S, when the interferometer is at rest relative to that frame. It follows from the isotropy of space relative to S

[17] As always, we have to be careful with the obvious! A very contrived model of electrodynamics involving anisotropic light propagation has recently been shown to be consistent with the null MM result when the orientation dependence of lengths of rigid bodies—itself a prediction of the theory—is taken into account. See Lämmerzahl and Haugan (2001).

that such rest lengths do not depend on the orientation of the interferometer. It is further taken (following assumptions 1 and 2 in section 2.4) that rest lengths are invariant, i.e. $L^o_{A(B)} = L'_{A(B)}$, and that $L'_{A(B)}$ likewise does not depend on the orientation of the interferometer. Now the fringe shift resulting from the rotation is proportional to $\Delta - \Delta^{rot}$, and for this quantity to be strictly zero, it must be the case that

$$\mathcal{C}_\perp = \gamma \mathcal{C}_\parallel. \tag{4.9}$$

Actually, an explanation of the null MM result ensues if the dimensions relative to S of the stone block underpinning the interferometer undergo an anisotropic change as a result of motion, consistent up to second order with (4.9), and hence with $\mathcal{C}_\perp \sim (1 + v^2/2c^2)\mathcal{C}_\parallel$. But using (4.3, 4.4, 4.9) we obtain the *MM-transformations*:

$$x' = k\gamma(x - vt) \tag{4.10}$$

$$y' = ky \tag{4.11}$$

$$z' = kz \tag{4.12}$$

$$t' = k(\gamma c/c')(t - vx/c^2), \tag{4.13}$$

where $|v| < c$, and putting $k \equiv 1/\mathcal{C}_\perp$.[18] The temporal transformation (4.13) can be rewritten as

$$t' = (\gamma^2/\mathcal{D})(t - vx/c^2), \tag{4.14}$$

in which form it is clearer that the MM experiment (in its original form, but see below) does not constrain the time dilation factor $\mathcal{D}$. Let's examine the main points of significance concerning what has been established so far.

(i) The value of the dimensionless factor $k = k(v)$ is not fixed by convention, but by experiment: it affects the measurable degree to which the shape of rigid objects is affected by motion relative to frame S. (Note that unless k is an even function of v, the shape effects will be anisotropic.) Indeed, the motion-induced distortion or deformation associated with (4.9) is of course more general than the

[18] The MM-transformations were derived in Brown and Maia (1993) and Brown (2001), and are consistent with the classic analysis by Robertson (1949) of the experimental basis of relativistic kinematics.

purely longitudinal contraction we are familiar with in special relativity (in which $\mathcal{C}_\perp = 1$). The deformation may actually involve expansion and/or contraction effects, but it cannot vanish *in toto*. Regrettably, very few textbook treatments of the MM experiment recognize anything other than a purely longitudinal contraction as an explanation of the null result.[19]

(ii) No appeal was made in the derivation of the MM-transformations to the relativity principle. Indeed, in general the deformation effect defined for motion relative to an *arbitrary* frame may differ from that defined relative to S.

(iii) The transformations (4.10), (4.13) are in general incompatible with the principle of Reciprocity (see section 2.4). It is easy to show that reciprocity holds only if $\mathcal{D} = \gamma$, or equivalently if $c = c'$. But since the synchrony convention has already been fixed in both frames, reciprocity is now an empirical issue.

(iv) It can easily be seen from (2.25-6) that even if we choose the synchrony convention in the lab frame S' in such a fashion that $\alpha = 0$ (which we must do if we want Galilean kinematics to emerge and which does not affect the length-change factors defined relative to S), the result $\mathcal{C}_\perp \neq \mathcal{C}_\parallel$ is inconsistent with the Galilean rule of transformation of velocities:[20]

$$u'_{x'} = u_x - v \tag{4.15}$$

$$u'_{y'(z')} = u_{y(z)}. \tag{4.16}$$

As we mentioned at the end of the previous section, this fact is in conflict with a common textbook 'solution' of the Michelson–Morley conundrum.

In this section we have looked at the implication of the MM experiment from a modern, kinematical perspective. Let us return now to history.

4.3 FITZGERALD AND HEAVISIDE

In May of 1889, the American journal *Science* published a brief letter by the prominent Irish physicist George Francis FitzGerald (1851–1901),[21] containing a sensational suggestion: 'almost the only hypothesis' capable of reconciling the 'wonderfully delicate' MM experiment with the apparent fact that the earth

[19] The only exception I am aware of amongst modern textbooks is Mills (1994). As we see below, both Lorentz and (almost certainly) FitzGerald were aware that strict longitudinal contraction was not required, as well as a handful of commentators on Lorentz.

[20] See Melchor (1988) and Brown (2001).

[21] For good accounts of FitzGerald's character and life, see Lodge (1905) and Coey (2000). It is too bad that a full scientific biography worthy of this man, one of the great Maxwellians, and who by all accounts was revered by his colleagues and friends, has never been written.

dragged a negligible amount of ether at its surface was that . . . 'the length of material bodies changes, according as they are moving through the ether or across it, by an amount depending on the square of the ratio of their velocities to that of light.'[22]

FitzGerald's letter was to remain virtually unknown until the historian Stephen Brush drew attention to it in 1967.[23] FitzGerald himself never checked to see if it had appeared in print, and when his friend Joseph Larmor edited FitzGerald's collected work a year after his untimely death in 1901,[24] Larmor was unaware of its existence. Nowhere else did FitzGerald publish the idea that probably most made his name in physics,[25] and yet by the time of his death it was widely accepted at least in the European physics community. The reason for its fame was essentially twofold. It was mentioned in papers on optics by another of FitzGerald's friends and colleagues, Oliver Lodge, in 1892 and 1893, and H. A. Lorentz independently arrived at the same idea in 1892. Lorentz promulgated it more energetically than FitzGerald while recognizing the latter's priority. Joseph Larmor, following Lorentz, would adopt the idea in his 1900 book *Aether and Matter*. But FitzGerald would not find his idea easy to sell for some time. In 1884, when he mentioned it to R. T. Glazebrook and J. J. Thomson while visiting Cambridge on examining duties, his listeners found it—in Glazebrook's words—'the brilliant baseless guess of an Irish genius'.[26]

The few historical studies of this episode have not done it justice. Alfred Bork in 1966 was struck by the vagueness of the formulations of the hypothesis by both FitzGerald and Lodge: he noted that they 'do not state just what contraction is involved, in terms of mathematical details'.[27] On the other hand, Bruce Hunt claimed in 1988 that when FitzGerald first voiced the deformation hypothesis during a visit to Lodge's Liverpool home in 1889: 'there is no reason to think that the idea that flashed on him in Lodge's study involved anything other than a simple [longitudinal] contraction.'[28]

In my opinion, there is every reason to think FitzGerald was *not* thinking specifically of simple contraction—a term, incidentally, that he never used. Not only is longitudinal contraction not required to explain the MM result as we have seen above (and as Lorentz knew perfectly well), but Lodge's 1893 account of

[22] FitzGerald (1889). [23] Brush (1984). [24] Larmor (1902).

[25] FitzGerald did elsewhere endorse the idea in print when proposed by Larmor, but modestly without claiming priority; see FitzGerald (1900*b*) and FitzGerald (1900*a*).

[26] Glazebrook (1928). Lodge would proudly write in 1909 that the deformation hypothesis was born while FitzGerald was 'sitting in my study and discussing the matter with me. The suggestion bore the impress of truth from the first.' In his account of the episode, Silberstein (1914) wrote of Lodge: 'Happy are those who are gifted with that immediate feeling for "truth".' Alas, the feeling on this occasion may have been exaggerated. FitzGerald pointed out to Lorentz in 1894 that Lodge only mentioned the deformation hypothesis in his 1892, 1893 papers as a result of 'reiterated positiveness' on FitzGerald's part. It is also pretty clear from the papers themselves that Lodge was initially far from convinced of the validity of the hypothesis. For more details, see Brown (2001), section V.

[27] Bork (1966). [28] Hunt (1988).

the hypothesis explicitly mentions the possibility that 'the length and breadth of Michelson's stone block were differently affected'.[29] Nor did Lorentz attribute a strict contraction effect to FitzGerald; he wrote in 1899 in respect of the MM experiment: 'In order to explain the negative result of this experiment FITZGERALD and myself have supposed that, in consequence of the translation, the dimensions of the solid bodies serving to support the optical apparatus, are altered in a certain ratio.'[30]

In fact, W. M. Hicks in 1902 and Wolfgang Pauli as late as 1921 gave similar readings of the FitzGeraldian hypothesis. It is more recently that commentators have, almost without exception, incorrectly attributed a strict contraction hypothesis to FitzGerald.[31] The situation vis-à-vis Lorentz is better, but not much, as we shall see.

But the important question is: what could have made FitzGerald come up with such a strange and unlikely deformation hypothesis? His 1889 letter in *Science* provides part of the answer.

We know that electric forces are affected by the motion of electrified bodies relative to the ether and it seems a not improbable supposition that the molecular forces are affected by the motion and that the size of the body alters consequently.

The remaining piece in the puzzle was provided by Bruce Hunt in 1988.[32] Hunt pointed out that some months before he wrote the letter to *Science*, FitzGerald had been in communication with Oliver Heaviside (1850–1925), the remarkable autodidact who is credited with the familiar vector form of Maxwell's equations and by the turn of the century was widely regarded as one of the authorities on electrodynamics. The correspondence concerned Heaviside's 1888 analysis of the distortion of the electromagnetic field of a charged body resulting from its motion relative to the ether.[33] We need to pause to examine this development.

There are two ways to derive the electric and magnetic fields produced by charges in uniform motion relative to some inertial frame S, as textbooks on electromagnetism will testify. The easier way is to consider the relatively simple solution of Maxwell's equations associated with the rest frame of the charge—in which the electric field has spherical symmetry—and then use the known transformation rules for the components of the electric and magnetic fields under Lorentz transformations to derive the field strengths in the original frame S.[34] The second way is to solve Maxwell's equations in S for a moving charge using the so-called Liénard–Wiechert potentials, and then omitting the contributions

[29] Later Lodge would reminisce that FitzGerald, at the 1889 home meeting with Lodge, accepted his suggestion that the effect of motion might be a volume-preserving sheer distortion. In his 1966 study Bork found this astounding, and suggested it is likely that Lodge was projecting a view of his own. But it is entirely possible FitzGerald accepted Lodge's deformation as one of the possibilities consistent with the MM null result. See Brown (2001), section V.

[30] Lorentz (1899).

[31] The only exception I am aware of is Capria and Pambianco (1994); see Brown (2001), section V.

[32] Hunt (1988).

[33] Heaviside (1888).

[34] It is tempting to call this the modern way, but Larmor had used it before 1900!

to the field strengths that are due to any acceleration of the charge. Either way, the electric field is found to take the form

$$\mathbf{E} = \frac{(q\mathbf{r}/r^2)(1 - v^2/c^2)}{(1 - v^2 \sin^2\theta/c^2)^{3/2}}, \tag{4.17}$$

where **E** is evalutated at a point with displacement **r** from the centre of the charged body and θ is the angle between **r** and the direction of motion of the charge. The terms in (4.17) involving the charge's speed v give rise to a distortion away from spherical symmetry. Indeed, the surface of equipotential forms an oblate spheriod, now known in the literature as a *Heaviside ellipsoid*, whose principle axes have ratios γ^{-1}:1:1.

The reason for this terminology is that Heaviside arrived at the formula (4.17) (as well as the correct expression for the magnetic field) in the 1888 work that was soon to come to FitzGerald's attention. Needless to say, Heaviside did not use the easier route above, because neither the formal notion nor meaning of the Lorentz covariance of Maxwell's equations was known in 1888. It is interesting that Heaviside gives no indication as to how he got (4.17), but simply argues that it has all the properties that the desired solution of Maxwell's equations would be expected to have![35]

In a letter written to Heaviside in January 1889, FitzGerald made the remarkable suggestion that the Heaviside distortion result might be applied 'to a theory of the forces between molecules' in a rigid body.[36] The implication of this suggestion is obvious: the shape of a body moving through the ether will be deformed, precisely as FitzGerald stated in his *Science* letter published later in the same year. Could FitzGerald then have *predicted* the null outcome of the MM experiment had Heaviside arrived at his distortion result before 1887?

[35] For a recent derivation in the spirit of Heaviside's approach, see Dmitriyev (2002). Heaviside made one mistake. Being concerned with the field associated with a moving point charge, Heaviside thought the effect would be the same as that produced by charge smeared non-uniformly on a perfectly conducting moving sphere. After all, a uniform distribution of charge on a stationary conducting sphere gives the same radial field lines as a stationary point charge. In private correspondence in 1892, G. Searle pointed out to him that the equilibrium distribution of charge associated with the Heaviside field distortion would not be over a sphere but over an ellipsoid, and it was Searle who in later papers of 1896 and 1897 introduced the term 'Heaviside ellipsoid'; see Searle (1896). For a treatment of this issue, see Redžić (1992) and the references therein. Recently, Redžić has argued that there is a tension between the relativity principle holding for Maxwell's equations and Searle's result, which had it been recognized by the ether theorists might have led to the notion of length contraction independently of the Michelson–Morley experiment; see Redžić (2004). But it is hard to imagine the relativity principle being used in this way, at least prior to 1900.

[36] The letter contains another remarkable suggestion, *viz.* that the speed of light might be a limiting speed. Was this the first time this notion was advanced? See also in this connection Bell (1992).

It is doubtful. Despite some claims to the contrary, it is highly questionable whether prior to 1889 FitzGerald thought, as Larmor would some years later, that intermolecular forces are electromagnetic in origin.[37] (Lorentz, certainly, was very non-commital on this issue until well into the twentieth century.) But as part of a response to the MM null result, FitzGerald's deformation hypothesis is, as he says in his *Science* letter, 'not improbable' in the light of the Heaviside result. Shape deformation produced by motion is far from the proverbial riddle wrapped in a mystery inside an enigma. And there is another feature of FitzGerald's argument that needs to be emphasized. For Heaviside and FitzGerald, it is motion of charges *relative to the ether* that causes the distortion of the electric field. But how does the ether achieve this feat? Very simply—*by way of Maxwell's equations*. Throw away the ether, and simply assume Maxwell's equations hold relative to some inertial frame S. According to FitzGerald's reasoning, it still is the case that moving bodies undergo shape deformation relative to S as long as the intermolecular force fields mimic electromagnetic ones. The ether, in and of itself, is doing virtually no work in FitzGerald's original argument!

4.4 LORENTZ

Hendrik Antoon Lorentz[38] (1853–1928) hit on essentially the same deformation hypothesis independently in 1892,[39] only later learning of FitzGerald's idea through Lodge's papers on optics. Lorentz's original suggestion was that the Michelson stone block undergoes a longitudinal contraction which up to second order coincides with the factor γ^{-1}, but he was aware that certain other changes in the dimensions of the body 'would answer the purpose equally well'. It was in his extensive 1895 essay (sometimes referred to briefly as the 'Versuch'[40]) on electromagnetic and optical phenomena associated with moving bodies that the point was brought out more systematically. Here, Lorentz introduced a longitudinal factor $C_{\|}$ written as $1+\delta$ and a transverse factor $C_{\perp}$ as $1+\epsilon$. He claimed

[37] For further discussion see Brown (2001), section VII.

[38] I cannot do justice here to the role played by Lorentz as the greatest of the ether theorists, nor to the scope of his achievements. It is enough to say that he and Joseph Larmor were responsible for establishing that 'atoms of electricity' are consistent with the Maxwellian programme, thus founding a clear-cut distinction between matter and the ether, or rather fields. If one has to encapsulate his personal greatness in a few words, the job is best left to Einstein.

> If we younger ones had known H. A. Lorentz only as a great luminary, our admiration and veneration for him would already have been of an extremely special kind. But what I feel when I think of H. A. Lorentz is not covered by a long way by that veneration alone. For me personally he meant more than all the others I have met in my life's journey. Just as he mastered physics and mathematical structures, so he mastered also himself,—with ease and perfect serenity. His quite extraordinary lack of human weaknesses never had a depressing influence on his fellow-men. Everyone felt his superiority; no one felt depressed by it. Einstein (1957), p. 8.

[39] Lorentz (1892). [40] Lorentz (1895).

that the null result requires

$$\epsilon - \delta \sim v^2/2c^2, \tag{4.18}$$

which is consistent to second order with (4.9). Lorentz stressed that the value of one of the quantities δ, ϵ remains undetermined. Indeed, he mentions the possibilities of joint values associated with pure transverse expansion, and with a combination of longitudinal contraction and transverse expansion. From the 1895 essay up to his 1906 New York lectures on the theory of the electron, Lorentz would repeatedly refer to his hypothesis as one dealing with the 'changes of dimension' of a moving solid body. Most commentators overlook this point, but unlike in the case of FitzGerald, there is a handful of writers who have appreciated the difference between length contraction and Lorentz's hypothesis.[41]

Another common misconception in the literature is that the FitzGerald–Lorentz 'contraction' hypothesis is somehow intrinsically different from the contraction effect Einstein deduced in 1905, because the former is more 'real' or 'absolute' or something of the sort. Consider the following claim found in a very useful recent collection of papers on the history and foundations of special relativity:

> This contraction hypothesis was regarded as an immediate forerunner of Einstein's theory of relativity. However, it should be emphasized that the concept of FitzGerald and Lorentz contradicts the fundamental principle of relativity because the FitzGerald–Lorentz contraction is absolute, while the Lorentz contraction in Einstein's theory is relative.[42]

What is presumably meant here is that the hypothesis was defined initially for rigid bodies moving relative to the rest frame S of the ether, and not relative to an arbitrary frame. This is true, but it was recognized well before Einstein was on the scene that this difference meant little in practice. S could be any inertial frame, as far as experiment could tell. The real difference was not the effect in itself, but the nature of its justification.

Again, like FitzGerald, Lorentz was not introducing a purely *ad hoc* mechanism to save the appearances. In a letter written to Einstein in 1915, and unearthed many years later by A. J. Kox,[43] Lorentz admitted that he had arrived at the idea of deformation shortly before he developed a dynamical plausibility argument. But develop one he did, and to his regret (expressed in the letter to Einstein) he did not emphasize it more from the outset.[44] In 1892 Lorentz had convinced himself that a collection of electrons held together in equilibrium[45] would undergo longitudinal contraction by the factor γ^{-1} when boosted relative to the ether. Lorentz argued

[41] See esp. Nercessian (1986, 1988). For details of other perceptive commentators on Lorentz, see Brown (2001), section VI.

[42] Hsu and Zhang (2001), p. 516.

[43] Lorentz (1915).

[44] See the epigraph at the beginning of this chapter.

[45] The notion deserves care; it has been noted in this connection by Janssen (1995) that it was already known that no such stable equilibrium based on purely electrostatic forces is strictly possible.

both in 1892 and 1895 that it was 'not far-fetched' to infer an ether-wind effect on the molecular forces similar to that associated with electrostatic ones. Although he flirted with a strong version of this correlation, according to which molecular forces exactly mimicked the electrostatic forces in this sense, Lorentz realized that there was 'no reason' for it. By 1899, he explicitly admitted that he had no means of determining the factor k in the possible values $C_{\|} = (k\gamma)^{-1}$ and $C_{\perp} = k^{-1}$.

Lorentz's 1892 argument concerning the collection of charges was followed in 1895 and 1899 by the first-order and second-order versions respectively of his so-called *theorem of corresponding states*, which were designed to show that no first- or second-order ether-wind effects would be discernible in experiments involving optics and electrodynamics. Note that it was in the context of the first-order version of the theorem that the notion of 'local time' appeared as an adjunct—with purely formal significance—to the Galilean coordinate transformations:

$$t' = t - vx/c^2. \tag{4.19}$$

It was not until 1908 that Lorentz would learn of, and immediately acknowledge, the priority of Woldemar Voigt who first introduced this transformation in 1887.[46]

The full logic of the theorem of corresponding states—at least by modern lights—is convoluted. Unlike FitzGerald's logic, it relies explicitly on the formal symmetry properties of Maxwell's equations; and in the second-order version of the theorem, the Lorentz coordinate transformations appear in their full glory. (In particular by 1904 Lorentz had largely convinced himself, if not Poincaré, that $k = 1$.)[47] Part of what makes the argument unfamiliar to modern eyes is that Lorentz's interpretation of these transformations is not the one Einstein would give them and which is standardly embraced today. Indeed, until Lorentz came to terms with Einstein's 1905 work, and somehow despite Poincaré's warnings, he continued to believe that the true coordinate transformations were the Galilean ones, and that the 'Lorentz' transformations (the terminology is due to Poincaré) were merely a useful formal device. These days, Lorentz's pre-1905 views are often discussed but seldom really understood. As we shall see later, J. S. Bell would courageously defend in 1976 what he called the 'Lorentzian pedagogy' in

[46] Voigt (1994). Voigt has been described, perhaps somewhat whimsically, as the 'unsung hero of special relativity' (see Hsu and Zhang (2001), 24). In his 1887 paper, Voigt showed that coordinate transformations exist—specifically the Lorentz transformations multiplied by γ^{-1}—which preserve the form of the wave equations in the elastic theory of light. The main object of the paper was to derive a new expression for the Doppler shift. It is unclear precisely how Voigt meant the transformations to be interpreted, or why the multiplicative factor γ^{-1} is what it is. For further discussion, see Kittel (1974). (Kittel cites a 1888 paper of Voigt's on the effect of motion on the propagation of light which predicts a 'perceptible effect', in violation of the null result in the MM experiment.)

[47] Part of Lorentz's argument for $k = 1$ contains the assumption that a moving electron, considered to be a small conducting sphere covered with a unit of charge, contracts longitudinally by the factor γ^{-1}. Now the nature of the force of cohesion on the surface of the electron that prevents it from disintegrating under the repulsive Coulomb force between different charge elements on the surface was completely unknown, and Lorentz's assumption had no clear justification.

teaching special relativity, but he showed little familiarity with the complexities of Lorentz's reasoning.

I will not go through these complexities,[48] but concentrate on the bottom line of Lorentz's argument. Imagine a system of electric and magnetic fields and charged particles, whose configuration over time, call it A, is a solution of Maxwell's equations. (It goes without saying that these equations are taken to hold in the ether rest frame S.) Then there exists another solution associated with a distinct configuration B, whose description in terms of the Lorentz-transformed (primed) coordinates is the same as configuration A relative to S—the 'corresponding state' in S'. It is important to realize that the Lorentz transformations involve field components as well as coordinates, and that both depend on the elusive scale factor k mentioned above. Now the fact that these primed coordinates and field components are in part 'fictitious', i.e. not judged to have direct operational significance, is *not* of crucial import. What is critical is the claim that configuration B represents the result of boosting configuration A, i.e. accelerating the system to the new state of motion relative to the rest frame S. If it is then *further* assumed that this very configuration change holds for *all* the equipment in the laboratory, including say the stone block in the Michelson interferometer, then no effects of ether drift are discernible in optical or electromagnetic experiments, at least to second order.[49]

Michel Janssen has highlighted the importance of this combined assumption regarding the effects of an active boost and called it the 'generalized contraction hypothesis'. As Janssen correctly asserted, it is a consequence of the claim that *all* the fundamental laws of physics, and not just those of electromagnetism, are Lorentz covariant.[50] (But is this entirely obvious? It surely depends to some extent on the details of the accelerative process, and we will return to this issue later.)[51] No such claim was made by Lorentz prior to 1905; nor is it obvious that Lorentz recognized that even the restricted version of the boost hypothesis concerning purely electrodynamical systems is a consequence of the symmetries of Maxwell's equations.

[48] Happily, excellent critical reconstructions of Lorentz's theorem of corresponding states, and its evolution from 1895 to 1899, are available; the most systematic I am aware of is found in the work of Michel Janssen. See Janssen (1995)—an example of history of physics at its best—and the briefer review in Janssen and Stachel (2003); the study by Rynasiewicz (1988) is also recommended. Note that Janssen also showed (see Janssen (1995, 2003)) that the null result obtained in the second-order ether-drift experiment of Trouton and Noble involving the measurement of torque on rotating capacitors could not be explained by appeal to the FitzGerald–Lorentz deformation hypothesis alone, even in its restricted contraction form, but actually requires appeal to Einstein's 1905 identification of energy and mass.

[49] This is something of an oversimplification. Lorentz exploited the fact that most optical experiments involve observation of a pattern of brightness and darkness, and argued that corresponding states produce identical patterns of this kind. But Lorentz's argument only holds, Janssen has claimed, if the patterns are *stationary*; see Janssen (1995).

[50] Janssen (1995), fn. 60, p. 205.

[51] See section 7.5.1 below.

This brings us to the little-known aspect of Lorentz's thinking that was mentioned in Chapter 1 and deserves another look. In 1889, the reader will recall, Lorentz commented on the claim of a year earlier made by the French physicist Alfred Liénard (1869–1958)—he of the Liénard–Wiechert potentials mentioned in the previous section—that in a variation of the MM experiment in which the arms of the interferometer contain a transparent medium, FitzGerald–Lorentz deformation would be insufficient to predict a null result. It is remarkable that instead of seizing on the possibility of a novel second-order ether-wind experiment uncovering what the MM experiment failed to, Lorentz expected (as did Liénard) that the result would still be null, and went about trying to see what could cause this. His conclusion: besides the change in dimensions of the apparatus, the period of oscillation of light waves leaving the moving source must be γ/k times the period obtaining when the source is at rest relative to the ether. Indeed, this dilation effect is a consequence of the second-order version of Lorentz's theorem of corresponding states, as he pointed out in 1899.[52]

The almost universal view that Lorentz had, prior to Einstein's work, no appreciation of the breakdown of the Newtonian nature of time may be due to a failure to distinguish between the two distinct components built into the Lorentz transformation for time, namely the dilation factor ($\mathcal{D} = \gamma$) and the relativity of simultaneity factor ($\alpha = v/c^2$). Lorentz consistently failed to understand the operational significance of his notion of 'local' time, which is connected with the latter. He did however have an intimation of time dilation in 1899, but inevitably there are caveats. First, it is very unlikely that Lorentz appreciated the universality of time dilation—that it would hold for *all* ideal clocks, whatever their construction. Second, the sense in which he is reading time dilation off the k-Lorentz transformations (which for him are, recall, nothing more than a convenient formal device) is quite different from the way we now read it off.

Lorentz noted that the theorem of corresponding states actually implies that the frequency of oscillating electrons in the light source is affected by motion of the source, and it is this fact that gives rise to the change in frequency of the emitted light. But Lorentz realized that the oscillating electrons only satisfy Newton's laws of motion if it is assumed that both their masses and the forces impressed on them depend on the electrons' velocity relative to the ether. The hypotheses in Lorentz's system were starting to pile up, and the spectre of *ad hoc*ness was increasingly hard to ignore (as Poincaré would complain).

It is interesting to ask how FitzGerald might in principle have dealt with the Liénard challenge. Recall that the FitzGerald–Heaviside approach to the original MM experiment makes no use of the symmetry properties of Maxwell's equations, or the existence of form-preserving coordinate transformations. What principle holding then in the ether frame S might one appeal to in order to account for the

[52] As far as I know, the first commentator to notice this intriguing development in Lorentz's thinking was Michel Janssen; see Janssen (1995) and Janssen and Stachel (2003).

expected new null result in the presence of transparent media? It is clear that the Fresnel drag coefficient, which it will be recalled determines the refractive index of the moving medium relative to the ether frame, will play a role. And of course *two* coefficients are needed: one defined for light moving parallel to the direction of motion, and one defined for the component of the light ray speed transverse to the direction of motion. Suppose at any rate that we provisionally accept the following exact relations:

$$c_{\parallel} = \frac{c/n + v}{1 + v/(nc)} \tag{4.20}$$

$$c_{\perp} = c/(n\gamma), \tag{4.21}$$

where $c_{\parallel}$ and $c_{\perp}$ are the one-way speeds for light propagating in the direction of the positive x- and y-axes inside the moving medium relative to S, and n is the refractive index of the medium when at rest relative to S. (The speeds for propagation in the opposite directions are obtained by replacing c with $-c$.) The usual Fresnel drag coefficients[53] follow from (4.20) and (4.21) up to first order. It is straightforward, but a little tedious, to show that the to-and-fro times T_A and T_B defined above are, to be consistent with (4.20) and (4.21), equal, with the value $2k\gamma L^o n/c$. Since rotation of the apparatus does not affect this equality, the null outcome is assured.

Where do the expressions (4.20) and (4.21) come from? The velocity transformation rules that follow from the Lorentz coordinate transformations. Pulling them out of the air is far from ideal, but it is no worse in meeting Liénard's challenge than explaining the null outcomes of first-order ether-wind experiments by way of the Fresnel drag coefficient, at least before a dynamical account of it is available (of which more later). Perhaps more importantly, the relations show that a MM-type experiment involving a transparent medium does not yield a non-null result, contrary to a recent claim in the literature.[54] Lorentz was right. But note that in the argument I have just presented, which is close in spirit to Lorentz's earlier 1892 and 1895 analyses of the MM experiment, dilation of the period of oscillation of the light source plays no direct role.[55]

[53] See, e.g., Larmor (1900), p. 49. As we shall see shortly, Larmor provided an interesting treatment of the modified MM experiment independently of Lorentz and possibly without being aware of Liénard's work.

[54] See Cahill (2004). It might be objected that since (4.20) and (4.21) are relativistic, it is no surprise that the outcome is consistent with the relativity principle. But the point here is that Cahill's derivation of a non-null result assumes that the light speed in the moving medium is c/n relative to S, which is consistent with neither the Galilean nor the Lorentz transformations.

[55] Since (4.20) and (4.21) follow from the relativistic transformation rules for longitudinal and tranverse velocities, time dilation (defined up to the unknown factor k) is unavoidable, but it does not figure in the argument.

4.5 LARMOR

In 1898, Joseph Larmor (1857–1942), an accomplished Irish mathematical physicist lecturing at St John's College, Cambridge, put together a series of his papers on the electron theory. The resulting extended essay won the Adams prize at Cambridge that year and in 1900 was published as a book under the title *Aether and Matter. A development of the dynamical relations of the aether to material systems.*[56] Larmor's work has received less attention from historians of relativity theory than Lorentz's.[57] Abraham Pais, in his acclaimed 1982 scientific biography of Einstein, wrote that

> ... [T]here is no doubt that he [Larmor] gave the Lorentz transformations and the resulting contraction argument before Lorentz independently did the same. It is a curious fact that neither in the correspondence between Larmor and Lorentz nor in Lorentz's papers is there any mention of this contribution by Larmor.[58]

There is much truth to these remarks[59], but Pais's additional claim that Larmor also provided 'the proof that one arrives at the FitzGerald–Lorentz contraction with the help of these transformations' does not bear scrutiny.

Larmor was, it seems, the first to introduce the Lorentz transformations, although in the ungainly form

$$x_1 = \epsilon^{1/2}x' \tag{4.22}$$

$$y_1 = y' \tag{4.23}$$

$$z_1 = z' \tag{4.24}$$

$$t_1 = \epsilon^{-1/2}t' - (v/C^2)\epsilon^{1/2}x', \tag{4.25}$$

where $\epsilon \equiv (1 - v^2/C^2)^{-1}$, C is the speed of light relative to the rest frame S of the ether, and the x' etc. are obtained from the S coordinates by way of the Galilean transformations. Larmor showed that Maxwell's equations for the free field are covariant under these transformations.[60] In the way in which he applied this result, Larmor's logic is very similar to that of Lorentz's theorem of

[56] Larmor (1900).

[57] But see Kittel (1974), Buchwald (1981), Warwick (1991), and particularly Darrigol (1994). Larmor would go on to hold the Lucasian Professorship of Mathematics at Cambridge; his successor in the post was P. A. M. Dirac.

[58] Pais (1982), p. 126.

[59] Note that Pais is not saying Larmor anticipated the FitzGerald–Lorentz hypothesis, which would be impossible given the relevant dates. Pais is aware that the latter hypothesis concerns a family of possible deformations, and his point was that Larmor settled on longitudinal contraction before Lorentz did.

[60] Larmor (1900), p. 174. Relevant sections of the book are reprinted in Hsu and Zhang (2001), pp. 27–41.

corresponding states. (Larmor was aware of of the first-order version of the theorem in Lorentz's 1895 *Versuch*, but could not have known Lorentz's 1899 second-order work.) The transformations (4.21–4.24) seem to have merely formal significance; indeed Larmor explicitly refers to the fields ('aetherial vectors') associated with a system of moving electrons (which Larmor takes to be singularities of the ether) 'referred to axes of (x', y', z') moving through the aether with uniform translatory velocity $(v, 0, 0)$'.[61] It seems that like Lorentz, Larmor still viewed the Galilean transformations as valid, but again like Lorentz, deemed that he had proved that a system of electrons is 'contracted in comparison with the fixed system in the ratio $\epsilon^{-1/2}$, or $1 - \frac{1}{2}v^2/C^2$, along the direction of its motion'.[62] (I say 'like Lorentz', but as we have seen Lorentz continued to struggle with the elusive value of the scale factor k for several more years. Larmor is perfectly aware of this scale ambiguity[63], and his his choice of $k = 1$ is entirely arbitrary.)[64] Finally, it is patent that the whole argument is approximate, and holds only to second order.[65]

It is important to recognize that the contraction of a system of electrons does not lead automatically to the (contraction version of the) FitzGerald–Lorentz hypothesis—a further assumption about forces of cohesion between the constituents of matter is needed. Larmor, just as much as FitzGerald and Lorentz, made this quite clear.

> We derive the result, correct to second order, that if the internal forces of a material system arise wholly from electrodynamic actions between the systems of electrons which constitute the atoms, then an effect of imparting to a steady material system a uniform velocity of translation is to produce a uniform contraction of the system in the direction of the motion, of amount $\epsilon^{-1/2}$, or $1 - \frac{1}{2}v^2/C^2$.

Larmor is no more reading length contraction off the coordinate transformations than Lorentz does; indeed earlier in *Aether and Matter*, in a discussion of the fields surrounding moving spherical conductors, he refers to 'a physical hypothesis presently to be discussed' which is that 'one effect of the motion is to actually cause a material system to shrink in this direction in the ratio $\epsilon^{-1/2}$.'[66] To repeat, Larmor's justification of this hypothesis is much closer to Lorentz's thinking than FitzGerald's, but in all cases appeal has to be made to the possibility that internal

[61] ibid. [62] op. cit., p. 175. Larmor denotes the velocity of radiation by C.

[63] op. cit., p. 176.

[64] This weakness in Larmor's logic is nicely treated by Darrigol (1994), pp. 330, who points out that Lorentz's 1904 solution to the k problem, such as it was, did not tally with Larmor's ideas about the point-like nature of the electron.

[65] Chapter XI of Larmor's 1900 essay, in which this argument is found, is an extension of the related first-order analysis found in chapter X; indeed the title of chapter XI is 'Moving material system: approximation carried to the second order'. This point was noted in Hsu and Zhang (2001), 41. Earlier interesting remarks on Larmor's eventual recognition of the exact nature of the Lorentz transformations—whose name he was happy to adopt in deference to Lorentz's 1892 work—as symmetries are found in Kittel (1974). [66] Larmor (1900), p. 155.

forces mimic electrostatic ones. Larmor is more gung-ho about this possibility than the ever-cautious Lorentz, reflecting the differing views the two men held on the nature of matter:

> It is to be observed that on the view being developed, in which atoms of matter are constituted of aggregations of electrons, the only actions between atoms are what may be described as electric forces. The electric character of the forces of chemical affinity was an accepted part of the chemical views of Davy, Berzelius, and Faraday; and more recent discussions . . . have invariably tended to the strengthening of that hypothesis.[67]

Indeed, from the point of view of the electrodynamics and optics of moving bodies, this seems to be the only issue on which Larmor goes beyond Lorentz. The debt Larmor owes to Lorentz and his 1895 theorem of corresponding states is reinforced in a paper Larmor published in 1904,[68] which offers another attempt to put the 'somewhat complex' argument for the contraction hypothesis (now explicitly associated with FitzGerald as well as Lorentz) which has been 'misunderstood'. Like *Aether and Matter*, this paper extends Lorentz's thinking to the second-order regime;[69] Larmor is still apparently unaware of Lorentz's independent work of 1899. Abraham Pais was right: it is one of the curiosities of this whole story how little Lorentz and Larmor were aware of each other's published work.

This leaves us with the question of how Larmor treats time in his theory. It is contentious. In 1937 Herbert Ives attributed time dilation to Larmor, and in the 1970s, the same attribution was urged by Charles Kittel and apparently independently by John S. Bell, after Wolfgang Rindler had scotched it.[70] Indeed, Bell referred to relativistic time dilation as the 'Larmor effect'.[71] How justified this nomenclature is depends on how you define time dilation.

It is true that in *Aether and Matter*, Larmor discusses the case of a pair of electrons of opposite signs moving in uniform circular orbits around each other, so slowly that radiation can be ignored. He shows, using arguments similar to those applied to a system of relatively static electrons, that when the pair is moving uniformly and rectilinearly through the ether, the orbit not only contracts longitudinally by

[67] Larmor (1900), p. 165. Darrigol (1994) argues that although Larmor's reasoning in relation to the theorem of corresponding states is very similar to Lorentz's, there was a crucial difference:

> Since Larmor conceived of matter as built out of point singularities of the ether, he could do without Lorentz' assumptions regarding the internal structure of electrons and the behaviour of molecular forces. (336)

Darrigol's 1994 study provides an illuminating account of the differences between the ether ontologies of Larmor and Lorentz (it appears only the latter adopted a truly dualist picture of ether and matter) and between their overall notions of explanation in physics.

[68] Larmor (1904).

[69] Note that Larmor in this paper (622) repeats Lorentz's point that although corresponding states involve distinct field configurations, the patterns of darkness are the same.

[70] See Ives (1937*b*), Kittel (1974), Bell (1976*a*), and Rindler (1970).

[71] Perhaps it is not a total irrelevance that both Larmor and Bell were Ulstermen!

the familiar ratio $1 - \frac{1}{2}v^2/C^2$, but the period of rotation will be changed in the ratio $1 + v^2/2C^2$.[72] In fact, the argument had already appeared in a paper Larmor published in 1887, as Kittel noted in 1974.[73] This dilation in the period is of course consistent to second order with the relativistic factor $\mathcal{D} = \gamma$. Again, Larmor would need to appeal to his internal forces hypothesis if he wanted to extend the dilation effect to macroscopic clocks of arbitrary construction. However, about this issue he remained silent, and one can only speculate as to how pervasive the 'Larmor effect' would have appeared to him. This is precisely Rindler's point: if it is not universal it is not true-blue time dilation. It seems churlish however to deny that Larmor had gained an important, if limited insight into time dilation, two years before Lorentz's strikingly similar and independent insight of 1899.

Recall that Lorentz was responding to the challenge laid down by Liénard, concerning a MM-type experiment with transparent media in the arms of the interferometer. It is remarkable that Larmor independently discusses precisely this possibility in chapter III of *Aether and Matter*, and demonstrates that using the then accepted formulae for the longitudinal and transverse Fresnel coefficient a non-null result is expected.[74] However, in chapter XI, Larmor uses corresponding states for the electromagnetic fields in order to calculate to second-order ray velocities in the moving medium, and now predicts not just a null result in the modified MM experiment, but 'the absence of effect of the Earth's motion in optical experiments, up to the second order of small quantities'. Note that once again, the ray velocities are not being read off the coordinate transformations (as was done at the end of the previous section above).

It is worth remarking that in both *Aether and Matter* and his 1904 paper, Larmor does acknowledge that the coordinate transformation (4.25) involves a change in the 'scale of time' in the ratio $\epsilon^{1/2}$. Note first that this factor is the *inverse* of the relativistic factor $\mathcal{D} = \gamma$. Second, it is anything but clear whether Larmor gave this scale change any general physical significance, despite the argument referred to above concerning orbiting electrons. In his 1904 paper, Larmor states enigmatically that the shrinkage in the scale of time 'being isotropic, is unrecognizable'.[75]

I give one last positive comparison with Lorentz. There seems to have been no appreciation on Larmor's part prior to 1905 that the vx/C^2 terms in the first- and

[72] Larmor (1900), p. 179.

[73] See Larmor (1897) and Kittel (1974). Kittel claims that this 'is the first historical statement of time dilation', p. 727.

[74] Larmor (1900), p. 50 gives general formulae valid up to second order for the transit times in each of the arms, for an arbitrary orientation of the instrument relative to the direction of the ether wind. Making use of both the Fresnel drag coefficient and the longitudinal contraction effect, it is easy to show that these formulae lead to a delay time (compare (4.5) above) $\Delta = (2nL^0v^2/c^3)(1 - 1/n^2)$ before rotation of the apparatus. Larmor warns however that 'according to the general molecular theory to be explained later [Chapter XI], it [the outcome of the MM experiment] will always be null'.

[75] Larmor (1904), p. 624.

second-order versions of the temporal coordinate transformation—associated with Lorentz's 'local' time—is connected with the issue of how to synchronize clocks at rest relative to the moving system of charges, etc.

Let's end this brief survey of Larmor's views with mention of his 1929 comments on Einstein's 1905 work. Like many commentators, Larmor mistakenly read Einstein's light postulate as that of the invariance of the light-speed from one frame to another. (As we shall see, light-speed invariance follows from Einstein's postulates; it is not one of them.) At any rate, he found this assumption 'mysterious', and one that 'might well appear to be a pure paradox', until it is put into the context of the theorem of corresponding states. Until, that is, it is given a dynamical underpinning based on the structure of Maxwell's equations, despite 'masquerading in the language of kinematics'.[76]

4.6 POINCARÉ

Of all the *fin de siècle* trailblazers, the one that came closest to pre-empting Einstein is Henri Poincaré (1854–1912)—the man E. T. Bell called the 'Last Universalist'. Indeed, the claim that this giant of pure and applied mathematics co-discovered special relativity is not uncommon[77], and it is not hard to see why.

(i) Poincaré was the first to extend the relativity principle to optics and electrodynamics *exactly*.[78] Whereas Lorentz, in his theorem of corresponding states, had from 1899 effectively assumed this extension of the relativity principle up to second-order effects, Poincaré took it to hold for all orders.

(ii) Poincaré was the first to show that Maxwell's equations with source terms are strictly Lorentz covariant. Actually, he proved that they are covariant under what we shall call the *k-Lorentz transformations*:[79]

$$x' = k\gamma(x - vt) \tag{4.26}$$

$$y' = ky \tag{4.27}$$

$$z' = kz \tag{4.28}$$

$$t' = k\gamma(t - vx/c^2). \tag{4.29}$$

[76] Larmor (1929), p. 644.

[77] See, e.g., Zahar (1983, 1989).

[78] A careful account of Poincaré's treatment of the relativity principle is found in Paty (1994).

[79] The proof is found in Poincaré (1906), of which a summary is found in Poincaré (1905). These famous papers, completed in mid-1905, are both entitled 'The dynamics of the electron'. (Extensive references to discussions of these papers in the literature are given in Darrigol (1995), which contains an astute analysis of Poincaré's work on electrodynamics.) The full covariance group of Maxwell's equations was not recognized until the work of Cunningham (1910) and Bateman (1910), which showed that besides the k-Lorentz transformations the group contains elements corresponding to certain non-linear coordinate transformations. (Note that Voigt's 1887 transformations correspond to $k = \gamma^{-1}$.)

(These are the MM-transformations (4.10–4.13) with (two-way) light speed invariance: $c' = c$. Note that the transformations of the **E**, **B** fields pick up k^2 terms etc.)

(iii) Poincaré was the first to use the generalized relativity principle as a constraint on the form of the coordinate transformations. He recognized that the relativity principle implies that the transformations form a group, and in further appealing to spatial istotropy, he independently, and virtually simultaneously, came up with the same argument as Einstein to determine the unit value of the dimensionless scale factor k in the equations (4.21–4.24), something that Lorentz had never quite succeeded in doing convincingly.

(iv) Poincaré was the first to see the connection between Lorentz's 'local time' (4.19) and the issue of clock synchrony. In 1900, he not only pointed out that the form of local time results from the adoption in both frames S and S' of the rule for synchronizing clocks in which the one-way light speed is isotropic,[80] but he insisted that such a rule was but a convention, though a natural one.[81] (This so-called 'Einstein' convention for synchronizing distant clocks would be more aptly entitled the 'Poincaré' or 'Poincaré–Einstein' convention. I shall adopt the latter terminology henceforth.) It is fair to say that Poincaré was the first to understand the relativity of simultaneity, and the conventionality of distant simultaneity.

(v) Poincaré anticipated Minkowski's interpretation of the Lorentz transformations as a passive, rigid rotation within a four-dimensional pseudo-Euclidean space-time. He was also aware that the the electromagnetic potentials transform in the manner of what is now called a Minkowski 4-vector.

(vi) He anticipated the major results of relativistic dynamics (and in particular the relativistic relations between force, momentum and velocity), but not $E_0 = mc^2$ in its full generality.

Taking all of this on board, is not the onus on the sceptic? What *are* the grounds for denying Poincaré the title of co-discoverer of special relativity? Here are some considerations that bear on what is bound to be a contentious issue.

(a) Although Poincaré understood independently of Einstein how the Lorentz transformations give rise to non-Galilean transformation rules for velocities

[80] Poincaré (1900). Thus Poincaré anticipated the derivation of (4.3) found in section 4.2 above. As noted by Janssen (1995) p. 248, it is ironic that Lorentz was oblivious to Poincaré's clarification of the significance of local time, given that it appeared in a collection of papers celebrating the 25th anniverary of Lorentz's doctoral thesis! Inexplicably, Lorentz would only fully understand the relativity of simultaneity through reading Einstein's 1905 work.

[81] In his 1898 essay 'The Measure of Time', Poincaré had already clarified that the supposition that the speed of light is the same in all directions 'could never be verified directly by measurement', and that 'it furnishes us with a new rule for the investigation of simultaneity'. It is pretty clear from the essay that Poincaré views the adoption of this rule, or alternatively that of clock transport, as ultimately based on convenience in relation to the problem at hand (Poincaré (1898), §XII).

(indeed Poincaré derived the correct relativistic rules), it is not clear that he had a full appreciation of the modern operational significance attached to coordinate transformations. Although it is sometimes claimed[82] that Poincaré understood that the primed coordinates (part of Lorentz's 'auxiliary quantities') were simply the coordinates read off by rods and clocks stationary relative to the primed frame, he did not seem to understand the role played by the second-order terms in the transformation. (Note that the γ's do not appear in the velocity transformations.) Let me spell this out.

Compared with the cases of Lorentz and Larmor, it is *even less* clear that Poincaré understood either length contraction or time dilation to be a consequence of the coordinate transformations. Take length contraction first. In proving $k = 1$ for the k-Lorentz transformations in 1906, Poincaré at no point says that he has thereby shown that the deformation is indeed a longitudinal contraction. He doesn't seem to connect the issues at all. A similar state of affairs is observed in his 1905 treatment of the deformability of the moving electron. One of the main results of his 1906 paper 'On the dynamics of the electron'[83] was the demonstration that amongst the existing rival notions concerning the shape of the moving electron (assumed to take the form of a sphere at rest) only the longitudinal contraction hypothesis of Lorentz is consistent with the relativity postulate. Once again, the argument made no appeal to the form of the coordinate transformations even after Poincaré had shown $k = 1$. The claim made by Abraham Pais that 'the reduction of the FitzGerald–Lorentz contraction to a consequence of Lorentz transformations is a product of the nineteenth century' in the context of Lorentz's 1899 work has been justly criticized by Janssen.[84] The claim is equally doubtful in relation to Larmor and wholly inappropriate for Poincaré. Pais himself emphasized the fact that as late as 1908, Poincaré still did not regard length contraction as a consequence of the relativity principle and Einstein's light postulate (or something close to it).[85]

Now take time dilation. It was claimed by Rindler in 1970 that *Poincaré never recognized its existence*, at least prior to Einstein.[86] I have found nothing in Poincaré's writings which contradicts this claim. In particular, his interpretation of Lorentz's local time (4.19) never seemed to be cognizant of the fact that in the second-order theorem of corresponding states, and indeed in his own treatment of the Lorentz covariance of Maxwell's equations, a multiplicative γ appears in the transformation.[87] It is very striking that his 1902 analysis of the measure of time is a repetition of his 1898 analysis.

[82] See Darrigol (1995) and Janssen and Stachel (2003).

[83] Poincaré (1906); a summary of this paper is found in Poincaré (1905).

[84] Janssen (1995), p. 212. [85] Pais (1982). [86] Rindler (1970).

[87] In Pais (1982), p. 133, it is emphasized that Einstein must have been aware before 1905 of Poincaré's insistance that we lack of any intuition about the equality of two time intervals. The reader of Pais's admirable book would be forgiven for thinking that perhaps Einstein was steered by this towards time dilation, but this would be wrong. Poincaré's point is related to the problem of comparing two different intervals within the same frame.

Let us go back to Lorentz's initial notion of local time (4.19). If this really is the time read off by suitably synchronized clocks stationary in the S' frame—the view that is normally attributed to Poincaré—then we need to compare (4.19) with the general form of the linear time transformation given in (2.24). It will be seen that Lorentz's seemingly first-order local time actually has hidden in it a dilation factor given by $\mathcal{D} = \gamma^2$! Of course, (4.19) appeared in the context of a theory defined only to first order, but even if it were exact, it is not obvious Poincaré would have recognized the dilation. It may never have occurred to him to analyse the coordinate transformations in the way that was done in Chapter 2 above. Or even if he did, he may have felt that not enough was known to justify the two assumptions given in section 2.4 needed to connect length change and time-dilation factors with the form of the coordinate transformations corresponding to boosts.

(b) It is well known that Poincaré was unhappy with the piling up of 'complementary' hypotheses in Lorentz's explanation of the absence of any ether-wind effects up to second order, but what I find most striking is his attitude to the first of these: the FitzGerald–Lorentz deformation hypothesis. Poincaré famously referred to it (which he seems always to interpret in terms of pure contraction) as a *coup de pouce* (helping hand) that Nature provides to secure the null result of the MM experiment. He seems never to have recognized the dynamical plausibility argument Lorentz gave concerning the intermolecular forces within solid bodies and their possible mimicry of electrostatic forces. Indeed, although he saw in Lorentz's theory of the electron the best available account of the electrodynamics of moving bodies, he did not credit it with the resources needed to provide an adequate explanation of the unobservability of motion relative to the ether. What Poincaré was holding out for was *no less than a new theory of ether and matter*—something far more ambitious than what appeared in Einstein's 1905 relativity paper. It is true, as we saw in (iii) above, that like Einstein, Poincaré would use the relativity postulate to constrain the form of the coordinate transformations associated with boosts. But this was only a temporary, stop-gap measure. Poincaré would have been in complete agreement with Lorentz when the latter wrote 'Einstein simply postulates what we have deduced, with some difficulty and not altogether satisfactorily, from the fundamental equations of the electromagnetic field.'[88] Actually, Lorentz followed this statement by a concession to Einstein: 'By doing so, he may certainly take credit for making us see in the negative [i.e. null] result of experiments like those of Michelson, Rayleigh and Brace, not a fortuitous compensation of opposing effects but the manifestation of a general and fundamental principle'.[89]

But the principle in question is of course none other than Poincaré's relativity postulate! Poincaré was already perfectly aware that that the null result of the MM experiment, for instance, was a direct consequence of it, and for all he did

[88] Lorentz (1916), p. 230. [89] ibid.

in his 1905 paper, Einstein was not needed to drive the lesson home. Poincaré, to repeat, wanted much more, including ultimately an explanation of the relativity postulate itself. Ironically, one is reminded of Einstein's own alleged response to the appearance in 1952 of the deterministic, hidden-variable version of quantum theory due to David Bohm—too 'cheap'. Like Einstein half a decade later, Poincaré wanted new physics, not a reinterpretation or reorganization of existing notions.

4.7 THE ROLE OF THE ETHER PRIOR TO EINSTEIN

It is hard not to believe that a major factor in the widespread view that special relativity was the brainchild of Einstein and not of the *fin de siècle* trailblazers has to do with the perceived role of the ether in the work of FitzGerald, Lorentz, Larmor, and Poincaré. After all, Einstein in 1905 summarily declared the notion of the ether 'superfluous', and his predecessors didn't. But let us not be too hasty.

I mentioned earlier that the mechanical nature of the ether played virtually no role in FitzGerald's arguments for the change of dimensions in rigid bodies moving through it. In the case of Lorentz, he was content to abandon attempts to model the ether's constitution, and came to regard it merely as 'the bearer of electromagnetic phenomena', as well as providing a frame relative to which true simultaneity could be defined (but not detected).

> ...[W]hether there is an aether or not, electromagnetic fields certainly exist, and so also does the energy of electrical oscillations. If we do not like the name of 'aether', we must use another word as a peg to hang all these things upon. It is not certain whether 'space' can be so extended as to take care not only of the geometrical properties but also of the electric ones.
>
> One cannot deny to the bearer of these properties a certain substantiality, and if so, then one may, in all modesty, call true time the time measured by clocks which are fixed in this medium, and consider simultaneity as a primary concept.[90]

It is noteworthy that as late as 1922, when the above remarks were published, Lorentz still does not seem to be aware of the conventional nature of distant simultaneity even in the rest frame of the 'medium', whatever its nature. But more significantly for our present purposes, Lorentz was never able to picture the electromagnetic field as an entity in its own right, rather than as a property of some kind of substratum.

A similar picture was painted by Poincaré in 1902, a decade before his death. 'We know the origin of our belief in the ether. If light takes several years to reach us from a distant star, it is no longer on the star, nor is it on the earth. It must be somewhere, and supported, so to speak, by some material agency.'[91]

Perhaps the most curious aspect of Poincare's views regarding the ether had to do with his understanding of the Fresnel 'drag' coefficient, that *deus ex machina* which

[90] Lorentz (1922), pp. 210–11. [91] Poincaré (1952), p.169.

accounted for all the null results of first-order ether-wind experiments. Even after Lorentz had given in 1895 a dynamical derivation of the coefficient on the basis of an entirely undragged ether, Poincare continued to support the original Fresnellian interpretation according to which a fraction of the ether is caught up inside a moving transparent body (which fraction, embarrassingly, depending on the frequency of the light!). The celebrated (non-null) 1851 interference experiment performed by Fizeau which corroborated Fresnel's hypothesis seemed to show for Poincaré '. . . two different media penetrating each other, and yet being displaced with respect to each other. The ether is all but in our grasp.'[92]

At any rate, we tend to snigger today at the idea that physical fields need a substantial peg like the luminiferous ether to hang on. But in doing so we lose sight of the fact that Einstein himself and many others after 1905 fell foul of a similar prejudice. The view that the *space-time manifold* is a substratum or bedrock, whose point elements physical fields are properties of, is just the twentieth-century version of the ether hypothesis. This point was nicely expressed by John Earman:

> When relativity theory banished the ether, the space-time manifold M began to function as a kind of dematerialized ether needed to support the fields. In the ninetheenth century the electromagnetic field was construed as the state of a material medium, the luminiferous ether; on postrelativity theory it seems that the electromagnetic field, and indeed all physical fields, must be construed as states of M. In a modern, pure field-theoretic physics, M functions as the basic substance, that is, the basic object of predication.[93]

This was written in 1989, when most physicists and philosophers were still either unaware, or had failed to absorb the lessons, of the fact that Einstein had already found himself in 1915 having to abandon precisely this reification of the space-time manifold in order to save his generally covariant gravitational field equations from the bogey of underdetermination. (It is noteworthy that Earman himself would figure prominently in the later awakening of the philosophical community to the issue of underdetermination in GR.)[94]

At the end of the day, it is always possible to add for whatever reason the notion of a privileged frame to special relativity, as long as one accepts that it will remain unobservable. (It is sobering to recall that something just like this goes on in several of the main interpretations of relativistic quantum theory. In the so-called Copenhagen interpretation of quantum mechanics, the privileged frame is that relative to which the wave function collapses instantaneously at the end of a measurement process; in the standard construal of the de Broglie–Bohm theory interpretation, it is that relative to which the 'hidden' corpuscles or fields act on each other instantaneously.)[95] If this issue was all that separated the trailblazers

[92] Poincaré (1952), p. 170. Like Einstein, Poincare failed in 1905–6 to appreciate that the Fresnel drag coefficient is a consequence of the Lorentz transformations, or rather the velocity transformation that both men independently derived from them. [93] Earman (1989), p. 155.

[94] See Earman and Norton (1987).

[95] The issue as to whether quantum theory is incompatible with special relativity will be discussed in Appendix B.

from Einstein, we would hardly be justified in assigning clear-cut priority for authorship of special relativity to Einstein. But as we have seen in this chapter, the real situation was rather different. The full meaning of relativistic kinematics was simply not properly understood before Einstein. Nor was the 'theory of relativity' as Einstein articulated it in 1905 anticipated even in its programmatic form.

Let us leave the last word with Poincaré.

> Whether the ether exists or not matters little—let us leave that to the metaphysicians; what is essential for us is, that everything happens as if it existed, and that this hypothesis is found to be suitable for the explanation of phenomena. After all, have we any other reason for believing in the existence of material objects?[96]

Here Poincaré, writing in 1902, is espousing an anti-metaphysical view about the nature of reality that is very similar to Einstein's realism,[97] even if its application in this case seems non-Einsteinian. But he has a surprise in store for us: he goes on to assert that the existence of matter will never cease to be a convenient hypothesis, '. . . while some day, no doubt, the ether will be thrown aside as useless'.

[96] Poincaré (1952), pp. 211–12. As Darrigol (1995), p. 19 notes, these remarks by Poincaré date back to 1888.

[97] For a careful discussion of Einstein's notion of realism, see Fine (1986), chap. 6.

5

Einstein's Principle-theory Approach

> The principle of relativity is a principle that narrows the possibilities; it is not a model, just as the second law of thermodynamics is not a model.
>
> Albert Einstein[1]

5.1 EINSTEIN'S TEMPLATE: THERMODYNAMICS

How did Albert Einstein (1879–1955) arrive at his special theory of relativity? (There is a nice cartoon picturing the young Einstein with chalk in hand at the blackboard, writing his most famous equation after having crossed out $E = ma^2$, and $E = mb^2$.) Much work has been done by historians to trace out the steps in the intellectual journey that brought Einstein in his mid-twenties to write his *On the Electrodynamics of Moving Bodies*—the fourth of his five publications, including his dissertation, in the *annus mirabilis* of 1905. (The announcement of the equivalence of mass and rest energy was in the fifth.) I won't attempt to summarize this journey.[2] I want only to stress that it is impossible to understand Einstein's discovery (if that is the right word) of special relativity without taking on board the impact of the quantum in physics.[3]

Several months before he finished writing his paper on special relativity (henceforth SR), Einstein had written a revolutionary paper claiming that electromagnetic radiation has a granular structure. The suggestion that radiation was made of quanta—or photons as they would later be dubbed—was the basis of Einstein's

[1] This statement was made by Einstein in 1911 at a scientific meeting in Zurich; see Galison (2004), p. 268. In 1911 Einstein was still using 'principle of relativity' to mean theory of relativity; see the related remarks in Stachel (1995), reference [2], 323.

[2] For useful accounts of the journey see Stachel (2002*b*, 2005) and his editorial contribution 'Einstein on the theory or relativity' in Stachel *et al.* (1989), pp. 253–74, reprinted in Stachel (2002*a*), pp. 191–214. See also Rynasiewicz (2000*a*), Norton (2005), Pais (1982) chapters 6 and 7, and Miller (1981). Further related works by John Norton can be found at <www.pitt.edu/jdnorton/jdnorton.html>. Einstein's own 1912 reconstruction of his journey is found in Einstein (1995*b*), pp. 9–108. See also Sterrett (1998).

[3] The present chapter was completed before I discovered recent papers by Robert Rynasiewicz discussing the conceptual origins of Einstein's 1905 paper, which are not only in agreement with much of my account but supplement it in useful ways, particularly in relation to the dynamical part of the Einstein paper. See in particular Rynasiewicz (2005).

extraordinary treatment of the photoelectric effect in the same paper. This treatment would win its author the Nobel prize of 1921; acceptance of the photon by the physics community would take longer. But the immediate consequence of Einstein's commitment to the photon was to destabilize in his mind all the previous work on the electrodynamics of moving bodies.

All the work of the ether theorists was based on the assumption that Maxwellian electrodynamics is true. Poincaré was, as we have seen, waiting for the emergence of new physics of the interaction of the ether and ponderable matter, but even in his case it is not clear he expected any violation of Maxwell's equations. In the work of Lorentz, Larmor, and Poincaré, the Lorentz transformations make their appearance as symmetry transformations (whether considered approximate or otherwise) of these equations. But Maxwell's equations are strictly incompatible with the existence of the photon.

In his 1949 *Autobiographical Notes*, written when he was 67, Einstein was clear about the seismic implications of this conundrum.

> Reflections of this type [on the dual wave-particle nature of radiation] made it clear to me as long ago as shortly after 1900, i.e. shortly after Planck's trailblazing work, that neither mechanics nor electrodynamics could (except in limiting cases) claim exact validity. By and by I despaired of the possibility of discovering the true laws by means of constructive efforts based on known facts.[4]

Already in the *Notes*, Einstein had pointed out that the general validity of Newtonian mechanics came to grief with the success of the electrodynamics of Faraday and Maxwell, which led to Hertz's detection of electromagnetic waves: 'phenomena which by their very nature are detached from every ponderable matter'.[5] Later, he summarized the nature of Planck's 1900 derivation of his celebrated black-body radiation formula, in which quantization of absorption and emission of energy by the mechanical resonators is presupposed. Einstein noted that although this contradicted the received view, it was not immediately clear that electrodynamics, as opposed to mechanics, was violated. But now with the emergence of the light quantum, not even electrodynamics was sacrosanct. 'All my attempts . . . to adapt the theoretical foundation of physics to this [new type of] knowledge failed completely. It was if the ground had been pulled out from under one, with no firm foundation to be seen anywhere, upon which one could have built.'[6]

Earlier in the *Notes*, Einstein had sung the praises of classical thermodynamics, 'the only physical theory of universal content concerning which I am convinced that, within the framework of the applicability of its basic concepts, it will never be overthrown.'[7] Now, he explains how the very structure of the theory was influential in the search for a way out of the turn-of-the-century crisis in physics.

[4] Einstein (1969), p. 51, 53 [5] op. cit., p. 25 [6] op. cit., p. 45.

[7] op. cit., p. 33. Einstein wryly directs this confidence in thermodynamics to 'the special attention of those who are sceptics on principle'.

The longer and more despairingly I tried, the more I came to the conviction that only the discovery of a universal formal principle could lead us to assured results. The example I saw before me was thermodynamics. The general principle was there given in the theorem:[8] the laws of nature are such that it is impossible to construct a *perpetuum mobile* (of the first and second kind). How, then, could such a universal principle be found?[9]

Now it is very curious that Einstein here says 'principle' and not 'principles', given that (the kinematical part of) his 1905 paper rests on several postulates. I will return to this point later. Now that the connection with thermodynamics is revealed, it needs to be examined more closely.

5.2 THE PRINCIPLE *VS*. CONSTRUCTIVE THEORY DISTINCTION

Let us remind ourselves of what Einstein was attempting in his 1905 paper. The Maxwell–Lorentz theory is designed to account for the interaction between the ether and charged matter from the perspective of the ether rest-frame. Unfortunately, the earthbound laboratory, in which most tests of the theory are carried out, is moving relative to the ether. It would clearly represent a great simplification of the problem of 'the electrodynamics of moving bodies' if it could be shown that the equations of the theory are one and the same when expressed in relation to either of the ether or lab rest-frames. How should one go about proving this covariance if it is suspected that the fundamental Maxwell–Lorentz equations are themselves only of limited, i.e. statistical, validity?

Further insight into the nature of Einstein's thinking was revealed in 1919, in an article he wrote for the London *Times*, entitled 'What is the theory of relativity?'.[10] Here Einstein characterized SR as an example of a 'principle theory', methodologically akin to thermodynamics, as opposed to a 'constructive theory', akin to the kinetic theory of gases.

Most [theories in physics] are constructive. They attempt to build up a picture of the more complex phenomena out of the materials of a relatively simple formal scheme from which they start out. Thus, the kinetic theory of gases seeks to reduce mechanical, thermal, and diffusional processes to movements of molecules . . .

[Principle theories] employ the analytic, not the synthetic method. The elements which form their basis and starting point are not hypothetically constructed but empirically discovered ones, general characteristics of natural processes, principles that give rise to mathematically formulated criteria which the separate processes . . . have to satisfy . . . The theory of relativity belongs to the latter class.

[8] The word 'theorem' for 'Satze' in the translation by P. A. Schilpp is perhaps better rendered as 'sentence' or 'statement'. I thank Thomas Müller for discussion of this point.

[9] op. cit., p. 53.

[10] Einstein (1919).

If for some reason one is lacking the means of mechanically modelling the internal structure of the gas in a single-piston heat engine, say, one can always fall back on the laws of thermodynamics to shed light on the performance of that engine—laws which stipulate nothing about the structure of the working substances, or rather hold whatever that structure might be. The laws or principles of thermodynamics are phenomenological, based on a large body of empirical data; the first two laws can be expressed in terms of the impossibility of certain types of perpetual-motion machines. Could similar, well-established phenomenological laws be found, Einstein was asking, which would constrain the behaviour of moving rods and clocks without the need to know in detail what their internal dynamical structure is?

The methodological analogy between SR and thermodynamics was not a *post hoc* rationalization dreamt up by Einstein in 1919 and repeated in his *Autobiographical Notes*. He mentioned it on several earlier occasions. In a short paper of 1907 replying to a query of Ehrenfest on the deformable electron, he wrote:

> The principle of relativity, or, more exactly, the principle of relativity together with the principle of the constancy of velocity of light, is not to be conceived as a 'complete system', in fact, not as a system at all, but merely as a heuristic principle which, when considered by itself, contains only statements about rigid bodies, clocks, and light signals. It is only by requiring relations between otherwise seemingly unrelated laws that the theory of relativity provides additional statements. . . . we are not dealing here at all with a 'system' in which the individual laws are implicitly contained and from which they can be found by deduction alone, but only with a principle that (similar to the second law of the theory of heat) permits the reduction of certain laws to others.[11]

Einstein went on in this note to emphasize that even if the relativity theory is true, there is still ignorance as to the dynamics and kinematics (apart from parallel translation) of the rigid bodies over which charge is distributed in the theory of electrons.[12] In a letter to Sommerfeld of January 1908, Einstein wrote:

> So, first to the question of whether I consider the relativistic treatment of, e.g., the mechanics of electrons as definitive. No, certainly not. It seems to me too that a physical theory can be satisfactory only when it builds up its structures from *elementary* foundations. The theory of relativity is not more conclusively and absolutely satisfactory than, for example, classical thermodynamics was before Boltzmann had interpreted entropy as probability. If the Michelson–Morley experiment had not put us in the worst predicament, no one would have perceived the relativity theory as a (half) salvation. Besides, I believe that we are still far from having satisfactory elementary foundations for electrical and mechanical processes. I have come to this pessimistic view mainly as a result of endless, vain efforts to interpret the second universal constant in Planck's radiation law in an intuitive way.

[11] Einstein (1907*a*). The English translation is found in Document 44 in Einstein (1989).
[12] For discussion of the background to this episode, see Maltese and Orlando (1995), §2.

I even seriously doubt that it will be possible to maintain the general validity of Maxwell's equations for empty space.[13]

This passage is particularly interesting, because in making the connection with thermodynamics it highlights the price that is paid in adopting the principle theory approach to relativistic kinematics, and Einstein's unease with that price.[14] On a more upbeat note, in his 1917 text *Relativity*,[15] Einstein noted with satisfaction that in SR the explanation of Fizeau's 1851 experiment is achieved 'without the necessity of drawing on hypotheses as to the physical nature of the liquid', and that 'the contraction of moving bodies follows from the two fundamental principles of the theory, without the introduction of particular hypotheses'. He further noted that Lorentz, in correctly predicting the degree of deflection of high-velocity electrons (cathode- and beta-rays) in electromagnetic fields, needed an hypothesis concerning the deformability of the electron which was 'not justifiable by any electrodynamical facts', while the same predictions in SR are obtained 'without requiring any special hypothesis whatsoever as to the structure and behaviour of the electron'.

This is all within the spirit of his 1919 claim that SR is a principle theory. Perhaps the clearest statement by Einstein to this effect is the one written in 1955 to Born in which he expressed the significance of his 1905 discovery that 'the Lorentz transformation transcended its connection with Maxwell's equations and had to do with the nature of space and time in general.'

A further new result was that the Lorentz-invariance is a general condition for any physical theory. This was for me of particular importance because I had already found that Maxwell's theory did not account for the micro-structure of radiation and could not therefore have general validity.[16]

[13] Einstein (1995*a*).

[14] There is more than a hint of recognition of the limitations of his programme also in Einstein's 1911 remarks quoted in the epigraph at the beginning of this chapter. Such ambivalence about the nature of principle theories was not new:

[T]he variables of the science [of thermodynamics] range over macroscopic parameters such as temperature and volume. Whether the microphysics underlying these variables are motive atoms in the void or an imponderable fluid is largely irrelevant to this science. The developers of the theory both prided themselves on this fact and at the same time worried about it. Clausius, for instance was one of the first to speculate that heat consisted solely of the motion of particles (without an ether), for it made the equivalence of heat with mechanical work less surprising. However, as was common, he kept his ontological beliefs separate from his statement of the principles of thermodynamics, because he didn't wish to (in his words) 'taint' the latter with the speculative character of the former. (Callender (2001)).

[15] Einstein (1917).

[16] See Born *et al.* (1971), p. 248; see also Torretti (1983) note 8, pp. 292–3, and Zahar (1989), p. 101. For a discussion of Einstein's earliest doubts about Maxwell's equations, see Miller (1981), section 2.4.

5.3 EINSTEIN'S POSTULATES

5.3.1 The Relativity Principle

It appears that by 1899, Einstein had convinced himself that the relativity principle (RP) encompassed the laws of electrodynamics and hence optics.[17] In the beginning of his 1905 paper, he briefly discusses physical grounds for the 'conjecture that not only the phenomena of mechanics but also those of electrodynamics have no properties that correspond to the concept of absolute rest'.

Rather, the same laws of electrodynamics and optics will be valid for all coordinate systems in which the equations of mechanics hold good, as has already been shown for quantities of the first order. We shall raise this conjecture (whose content will hereafter be called 'the principle of relativity') to the status of a postulate.[18]

Einstein goes on to assert that the 'light ether' will prove to be 'superfluous' in his new account of the electrodynamics of moving bodies, in so far as no 'space at absolute rest' is required. Interestingly, he simply ignores the non-trivial issue (raised as we have seen by both Lorentz and Poincaré) as to whether the electromagnetic field can even be conceived of in the absence of the ether.

In section I.3, Einstein introduces yet a stronger claim: 'If two coordinate systems are in uniform parallel translational motion relative to each other, the laws according to which the states of a physical system change do not depend on which of the two systems these changes are related to.'

This principle applies simply to all fundamental laws in physics,[19] and the importance of this generalization—which Einstein may not initially have fully appreciated—will be seen below.

Note that these formulations of the RP makes no reference to the form of the coordinate transformations between the two frames in question. This is just as well! The whole point of the kinematical part of the 1905 paper is to derive the Lorentz transformations on the basis of a small number of phenomenological principles. If one of these principles, namely the RP, made appeal to the form of these transformations at the outset, the exercise would be circular. And the fact that Einstein's RP is silent on kinematics in this sense allows it to be (largely) identified with the relativity principle of Newton, as we saw in Chapter 3.

Einstein's formulation of the relativity principle has come under fire in recent times. It is sometimes regarded as too informal, or too imprecise; and the criterion that the fundamental equations of physical interactions take the same form in all inertial frames is certainly open to ambiguities. Michael Friedman, in particular,

[17] See Stachel (1995), p. 266.

[18] Einstein (1905*a*); the English translation is taken from Stachel (1998), pp. 123–60.

[19] The laws of sound propagation, for instance, are not fundamental; the mechanical medium in which the sound propagates acts as a natural symmetry-breaker.

has reminded us of a point (long ago raised by Kretschmann) that one can in principle construct dynamical theories incorporating a privileged inertial frame that are generally covariant (i.e. formulated in the tensor calculus associated with a four-dimensional space-time manifold). For Friedman the equations of such theories take the same form in *all* coordinate systems, not just inertial ones, and hence Einstein's rendition is 'too weak' to capture the content of the relativity principle. But this particular argument is flimsy. It is implicit in Einstein's definition that it is the 'simplest' form of the equations that is at stake, and this refers in the usual case of inertial coordinates in SR to the 3-tensor (mostly 3-vector) calculus. A theory may be formulated generally covariantly, but it doesn't follow that the equations are equally simple in all coordinate systems.[20] The attempts I am aware of that have been made in recent decades to give RP a more rigorous or precise formulation have invariably (a) misleadingly attributed different principles to Newton and Einstein and (b) involved some loss of physical insight.[21]

5.3.2 The Light Postulate

It is the ultimate irony that the paper which would spell the demise of the luminiferous ether had as one of its central postulates what Wolfgang Pauli aptly called (as we saw in Chapter 4) the 'true essence of the old aether point of view'.[22] Here is how it is presented in the 1905 paper: 'Every ray of light moves in the "rest" coordinate system with a definite velocity V, independently of whether this ray is emitted by a body at rest or in motion.'

Einstein made it clear from the beginning of his kinematical discussion that the 'rest' frame can be chosen arbitrarily from the infinity of inertial frames, the term 'rest' merely distinguishing it 'verbally' from all the others. However, this implicit appeal to the relativity principle at the outset does little to clarify the origins of the light postulate.

Throughout the 1905 paper, the physics being assumed to hold in the rest frame corresponds to what informed readers would expect to hold in the frame relative to which the ether is at rest. In a footnote to his fifth 1905 paper, on the inertia–energy connection, Einstein says in reference to his paper currently under discussion: 'The principle of the constancy of the velocity of light used there is of course contained in Maxwell's equations.'[23] In 1907, Einstein would write of the postulate:

> It is by no means self-evident that the assumption made here, which we will call 'the principle of the constancy of the velocity of light,' is actually realized in nature, but—at

[20] Indeed, Friedman (1983, p. 61) himself discusses the 'standard formulation' of a theory, which generally does not coincide with its generally covariant formulation, precisely because the theory may look simplest in a given family of coordinate systems.

[21] The attempt in Friedman (1983) is analysed in Sypel and Brown (1992); that of Norton (1989, 1993) and Wald (1984) is analysed in Brown and Sypel (1995) and Budden (1997*a*).

[22] Pauli (1981), p. 5.

[23] Einstein (1905*b*); an English translation is found in Stachel (1998), pp. 161–4.

least for a coordinate system in a certain state of motion—it is made plausible by the confirmation of the Lorentz theory, which is based on the assumption of an ether absolutely at rest, through experiment.[24]

Much later, Einstein would reiterate the point: 'Scientists owe their confidence in this proposition to the Maxwell–Lorentz theory of electrodynamics.'[25]

In assessing the Maxwell–Lorentz theory, Einstein was aware of some of the difficulties involved in constructing a reasonable theory of electrodynamics on the basis of the alternative *emission* theory of light.[26] And an important feature of Lorentz's 1895 theorem of corresponding states, and one cited by Einstein in his 1907 paper, was that it provided the first dynamical derivation of the Fresnel drag coefficient (shown by von Laue also in 1907 to be a direct consequence of the Einstein–Poincaré velocity transformations.) All this gives some indication as to why Einstein felt secure in postulating in 1905 the existence of a specific inertial frame relative to which the light-speed is independent of the source and isotropic. In effect then, Einstein was adopting, with nefarious intent, the Light Principle of the ether theorists which was introduced in section 4.2 above.[27] Thus the acronym LP can be used for both the etherial Light Principle or Einstein's light postulate. (It is worth recalling that in 1905 there was no direct empirical evidence supporting the LP, distinct from the empirical successes of the Maxwell–Lorentz theory generally. Today, it is known to hold to an accuracy of better than one part in 10^{11}.)[28]

It is often wrongly claimed that Einstein's light postulate is the stronger claim that the light speed is invariant across inertial frames. The advantage of his postulate as it stands is that it is logically independent of the RP. This meets an obvious desideratum in a semi-axiomatic derivation of the new kinematics of the type Einstein was constructing. The corresponding behaviour of light relative to any 'moving' frame was explicitly taken by Einstein in section 3 of the 1905 paper to follow from the conjunction of the light postulate and the RP. This would be reiterated in his later work. In 1921 Einstein wrote:

[24] Einstein (1907*b*). It should be noted that Einstein here equates the 'principle of the constancy of the velocity of light' with the claim that *c* is a 'universal constant', a claim also made in the section 1.1 of the 1905 paper. The significance of Einstein's remarks on the light postulate in the 1907 paper are discussed in Miller (1981), p. 202; our understanding of the postulate and the role of the resting frame in it appears to differ from Miller's (see also his discussion on p. 165).

[25] Einstein (1950), p. 56.

[26] Emission theories continued to be defended in certain quarters after 1905, most famously, as we have seen, by Walter Ritz in 1908. A useful discussion of these theories is found in Pauli's great treatise on relativity; see Pauli (1981), Part I, Section 3. Einstein's own flirtation with an emission theory is discussed in Stachel (1982), pp. 51–2.

[27] Recent recognition of the etherial origins of the light postulate is found in Galison (2004), p. 17; earlier cases of such recognition are listed in Brown and Maia (1993).

[28] The history of direct tests, both terrestrial and astronomical is fascinating, but beyond our scope; see Fox (1962). Note that the most precise test to date has been provided by neutrino detection events associated with Supernova 1987A; see Brecher and Yun (1988). The 1960s saw a number of terrestrial tests, particularly by T. Alväger and collaborators; for a critical review of these experiments see Hsu and Zhang (2001), chap. 17.

The consequence of the Maxwell–Lorentz equations that in a vacuum light is propagated with the velocity *c*, at least with respect to a definite inertial system K, must therefore be regarded as proved. According to the special theory of relativity, we must assume the truth of this principle for every other inertial system.[29]

And in his *Autobiographical Notes* Einstein defines the principle of relativity as the 'independence of the laws (thus specially also of the law of the constancy of the light velocity) of the choice of the inertial system . . . '.[30]

The most remarkable feature of Einstein's light postulate is the fact that it seems at first sight antithetical to his own revolutionary notion of the light quantum. In 1905 it was far from clear to Einstein what sort of thing the light quantum precisely is, but it must have seemed closer in nature to a bullet than a wave. The fact that nonetheless Einstein adopted the LP over an emission theory of light is testimony to the sureness of his physical intuition in the midst of blooming, buzzing confusion.[31]

5.4 EINSTEIN'S DERIVATION OF THE LORENTZ TRANSFORMATIONS

Einstein's 1905 derivation of relativistic kinematics is fairly elementary from a mathematical point of view, but it is conceptually subtle. It is useful to break the logic into three steps; what follows is a reconstruction of the argument rather than an exact rendition of the original.

5.4.1 Clock Synchrony

Relative to the rest frame S it is assumed as part of the light postulate that the two-way or round-trip velocity of light *in vacuo* c is isotropic, and it seems natural to adopt the Poincaré–Einstein convention for synchronizing distant clocks, according to which the one-way speed is also isotropic, and hence also has the value c.[32] (Remarkably, there persist doubts in the literature as to whether this is a true convention or whether it is imposed by, say, the structure of space-time. I will return to this issue in the following chapter.) According to the relativity

[29] Einstein (1921).

[30] Einstein (1969), p. 57. Detailed analyses of the logic of the kinematical part of the 1905 paper are found in Williamson (1977) and Rynasiewicz (2000*a*). Torretti (1983, note 7, p. 295) also stresses Einstein's 'careful reasoning' in deriving the light velocity in the moving frame from the light postulate and the relativity principle. See also Brown and Maia (1993), from which most of the material in this subsection is taken.

[31] See Hoffmann (1982) p. 99, and Torretti (1983) p. 49. The role of light quanta in Einstein's 1905 relativity paper is discussed in Miller (1976). An enlightening study of the relationship between early relativity and early quantum theory is found in Nugaev (1988).

[32] Einstein would have made things clearer if he had formulated the light postulate in terms of the two-way and not the one-way velocity of light.

principle, or alternatively the Michelson–Morley experiment, the same considerations apply to the moving frame S'. Note that what motivates the adoption of the Poincaré–Einstein convention in this case is again the isotropy of the two-way velocity, not light-speed invariance *per se*. It will be recalled from section 4.2 above that the mere adoption of this convention in both frames constrains the relativity of simultaneity factor α in (2.24) to take the value v/c^2. (This point was appreciated in section 3 of the Kinematic Part of the 1905 paper.)

5.4.2 The *k*-Lorentz Transformations

Einstein, as mentioned above, used the combination of the RP and the light postulate to infer the invariance of the velocity of light c, and from this he derived, in a somewhat convoluted argument, the *k*-Lorentz transformations (4.26–4.29). In section 3 of the paper, he showed these to be consistent with the claim that a spherical light-wave front centred at the origin seen in relation to the rest system S will also appear spherical from the point of view of the moving frame S', so that

$$x^2 + y^2 + z^2 - c^2t^2 = 0 \iff x'^2 + y'^2 + z'^2 - c^2t'^2 = 0. \qquad (5.1)$$

Nowadays, the *k*-transformations are standardly derived directly from (5.1), and the details will not be repeated here.[33]

A few commentators have questioned, somewhat idiosyncratically, whether the invariance of c does in fact follow from Einstein's two main postulates. The bone of contention is the meaning of the RP, and in particular the issue as to whether form-invariance of equations also means the invariance of the numerical values of the fundamental physical constants that appear in them. It might be considered that Einstein's postulates imply merely that relative to any moving frame S', there is a 'constant' two-way light velocity c', independent of the speed of the source and isotropic. Various attempts have been made to see what else is needed in this case to obtain the *k*-Lorentz transformations, instead of what we have called earlier the MM-transformations (4.10–4.13).[34] It should not be overlooked, however, that if distinct inertial frames came equipped with distinct, constant light speeds, empirical discrimination between them would be possible, and it is clear that for Einstein at least such a possibility was incompatible with the RP.

5.4.3 RP and Isotropy

Einstein, like Poincaré, took it to follow from the relativity principle also that the coordinate transformations form a group. In particular this means that the

[33] The 1913 reprint of the 1905 paper has an appended note acknowledging that this route is simpler than that given originally. It is not entirely clear whether Einstein wrote the note, but he may at least have approved it. See Stachel (1998), pp. 160–1.

[34] See Brown and Maia (1993) and the further references therein.

transformations between *arbitrary* pairs of frames take the same functional form as that of the original transformations taking S into S'. Consider now, in the case of 1+1-dimensional space-time, the k-Lorentz transformations with the following value of k:

$$k = \left(\frac{c+v}{c-v}\right)^n \tag{5.2}$$

for any real number n. I have referred elsewhere to the resulting transformations as the *Bogoslovsky–Budden (BB) transformations*.[35] They can be shown to form a group, so they are consistent with the relativity principle.[36] They are clearly consistent with the invariance of the light velocity. So why are the BB-transformations not the basis of relativistic kinematics? (Note first that in the special case corresponding to the value $n = 1/2$, although there is a time dilation factor $\mathcal{D} \neq 1$, there is no clock retardation effect, or 'twin paradox'.[37] Second, the BB transformations are associated with Finslerian, rather than a pseudo-Riemannian space-time geometry.)

The simple reason is that k in (5.2) is not an even function of v, so that the length change and clock dilation factors will differ depending on whether they move at velocity v or $-v$. (This explains the absence of a twin effect: the moving twin undoes on the return part of the trip the dilation it suffers on the outgoing part.) There is clearly a clash with the assumption that space associated with the frame S is isotropic.

What the existence of the BB-transformations highlights is the importance of the assumption of spatial isotropy in the Einsteinian derivation of the Lorentz transformations. The simplest way now of obtaining $k = 1$ is as follows.

Consider the transverse transformation $z' = k(v)z$. Inverting algebraically, we obtain $z = [k(v)]^{-1}z'$. But according to the relativity principle, the inverse transformations must take the same form as the original ones: $z = k(-v)z'$. This is only possible if

$$k(-v)k(v) = 1. \tag{5.3}$$

But now note that the transverse length change factor defined for a rigid body moving relative to the frame S at velocity v relative to the positive x-axis is given

[35] Brown (1997). To the best of my knowledge, they first appeared in Bogoslovsky (1977), which established a connection with Finsler geometry. The special case corresponding to the value $n = 1/2$ was introduced independently in Brown (1990), and Budden (1992) rediscovered the BB-transformations in generalizing Brown's work. An elegant survey of the nature and lessons of these kinematics is found in Budden (1997*b*), in which, as in Bogoslovsky (1977), a treatment in 4-dimensional space-time is given.

[36] It is necessary for the transformations to form a group in order that the RP is satisfied, but arguably not sufficient. The MM transformations (4.10–4.13) can be shown to form a group when $k = 1$, but they are not consistent with the RP when $c' \neq c$; see Brown and Maia (1993).

[37] Brown (1990).

by $C_\perp = (k(v)\gamma)^{-1}$, and it should take the same value, given isotropy, when v is replaced by $-v$. This means that $k(v) = k(-v)$, so from (5.3) we obtain $k^2 = 1$, and $k = 1$ is the only solution consistent with the transformations reverting to the identity transformation in the limit $v \longrightarrow 0$.[38] The Lorentz transformations

$$x' = \gamma(x - vt) \tag{5.4}$$

$$y' = y \tag{5.5}$$

$$z' = z \tag{5.6}$$

$$t' = \gamma(t - vx/c^2), \tag{5.7}$$

where $v/c < 1$, have thus been derived. As we shall see in the next section, there is more to be said about the second application of the RP in this third step.

5.5 RODS AND CLOCKS

The introductory section of the 1905 paper ends with the following paragraph.

> Like all electrodynamics, the theory to be developed here is based on the kinematics of a rigid body, since the assertions of any such theory have to do with the relations among rigid bodies (coordinate systems), clocks, and electromagnetic processes. Insufficient regard for this circumstance is at the root of the difficulties with which the electrodynamics of moving bodies has to contend.

At first sight, this last statement is surprising. Did not Lorentz clearly introduce shape deformation of rigid bodies in his 1895 *Versuch*, which Einstein is known to have been familiar with? Einstein's point here seems to be that before one understands the meaning of the coordinate transformations, one must understand what a coordinate system is. The same point is repeated in his *Autobiographical Notes*: 'One had to understand clearly what the spatial co-ordinates and the temporal duration of events meant in physics.'[39]

At the beginning of the Kinematical Part of the paper, Einstein is clearly more concerned with time than space. Positions of bodies are simply 'defined' with the aid of rigid measuring rods provided in the 'rest' system. (Actually, this is the beginning of Einstein's lifelong struggle with the notion of spatial distance, which we alluded to in section 2.2.4.) The real emphasis is on *motion*, which as Einstein points out introduces the notion of spreading time through space.

[38] Poincaré had independently derived $k = 1$ in a letter to Lorentz written sometime between late 1904 and mid-1905, and in essentially the same manner. Poincaré showed that if k takes the form $(1 - v^2/c^2)^m$, where m is an arbitrary real number, then the group property of the k-Lorentz transformations requires that $m = 0$. (See Miller (1981), section 1.14.) By assuming k to be an even function of v, over and above the condition that $k \rightarrow 1$ when $v \rightarrow 0$, Poincaré seems to be have been aware that the issue of spatial isotropy is important. See also his derivation of $k = 1$ in Poincaré (1906).

[39] Einstein (1969), p. 55.

Unfortunately, Einstein opts for an excessively operational treatment of this issue, which arguably has misled generations of students of the theory (and not a few philosophers). The picture that is often advocated in textbooks of vast, tightly packed grids in space with an ideal clock sitting at each intersection of the grid lines is traceable back to Einstein's paper, where a meaningful notion of one-way uniform motion, and indeed that of light itself, presupposes distinct points of space equipped with synchronized clocks. Nothing could be further from the truth in typical experiments. I will take up this issue again in the following chapter, but it must be conceded here that Einstein was right to put so much emphasis on the stipulatory nature of the isotropy of the one-way speed of light, and the conventional nature of distant simultaneity. It was essential in understanding the relativity of simultaneity, and hence the compatibility of Einstein's apparently irreconcilable postulates.

In this respect, Einstein was doing little more than expanding on a theme that Poincaré had already introduced, as we have seen. Where Einstein goes well beyond the great mathematician is in his treatment of the coordinate transformations as encoding information about the behaviour of moving rigid rods and clocks. Einstein's treatment is far closer to the discussion on the physics of coordinate transformations given in section 2.4 above than was ever given before. In particular, *the extraction of the phenomena of length contraction and time dilation directly from the Lorentz transformations in section 4 of the 1905 paper is completely original.* It was in particular the first time that the phenomenon of time dilation was predicted in its full scope.[40] Note that the two overarching assumptions of our section 2.4—the universality and boostability assumptions—are implicit in Einstein's treatment. In 1910, he was more explicit: 'It should be noted that we will always implicitly assume that the fact of a measuring rod or clock being set in motion or brought back to rest does not change the length of the rod or the rate of the clock.'[41]

The difference with Poincaré's approach is particularly clear, oddly, in a place where both protagonists seemed to agree. I refer to the second application of

[40] As Pauli (1981, p. 13) wrote: 'While this consequence of of the Lorentz transformations [time dilation] was already implicitly contained in Lorentz's and Poincaré's results, it received its first clear statement only by Einstein.'

See also Rindler (1970) in this connection. It is remarkable that in his detailed account of Einstein's 1905 reasoning about time, Galison (2004, chap. 5) does not mention the discovery of time dilation.

[41] Einstein (1910). The importance of this assumption was stressed by Born.

> . . . it is assumed as self-evident that a measuring rod which is brought into one system of reference S and then into another S' under exactly the same physical conditions would represent the same length in each. . . . Exactly the same would be postulated for clocks . . . We might call this tacit assumption of Einstein's theory the 'principle of the physical identity of the units of measure' . . . This is the feature of Einstein's theory by which it rises above the standpoint of a mere convention and asserts definite properties of real bodies. Born (1965) pp. 251–2.

As was argued in section 2.4, this is the feature that allows the coordinate transformations to encode the phenomena of length change and time dilation for moving rods and clocks respectively. (I am grateful to Antony Valentini for the Einstein and Born quotations; see Valentini (2005).)

the relativity principle in Einstein's argument, designed to guarantee the group property of the coordinate transformations. For Poincaré, this aspect of the transformations is intimately connected with their role as (a subgroup of) the covariance group of Maxwell's equations. For Einstein, who for reasons spelt out in section 5.1, cannot appeal to Maxwellian electrodynamics in the same way, the group property must be justified by appeal to what the transformations *mean in themselves*.

This is how I take it that the argument—which Einstein does not fully spell out—ought to go.[42] Suppose that the coordinate transformations between frames S and S' are different in form from their inverse, for instance. We expect in this case either the length contraction factor or the time dilation factor (if any), or both, to differ when measured relative to S and when measured relative to S'. And this would imply a violation of the relativity principle. Specifically, it would be inconsistent with the claim that the dynamics of all the fundamental non-gravitational interactions which play a role in the cohesion of these bodies satisfy the relativity principle. Thus the *dynamical* relativity principle constrains the form of the *kinematical* transformations, because such kinematics encode the universal dynamical behaviour of rods and clocks in motion. One can understand why, as we saw earlier, someone like Larmor would say that Einstein's 1905 reasoning is ultimately based on dynamical considerations 'masquerading in the language of kinematics'.

5.6 THE EXPERIMENTAL EVIDENCE FOR THE LORENTZ TRANSFORMATIONS

5.6.1 The 1932 Kennedy–Thorndike Experiment

We have already met the MM-transformations (4.10–4.13) that are a consequence of the null result of the Michelson–Morley experiment, where S is the frame relative to which the Light Principle is valid. It was stressed that the form of the temporal transformation (4.13) depends on the adoption of the Poincaré–Einstein convention for synchronizing clocks in both frames. In 1979, L. Brillet and J. L. Hall improved the accuracy of the MM experiment by a factor of 4000, using a helium-neon laser mounted on a rotary platform.[43] More recently, groups at Stanford University, Humboldt University in Berlin, and the Observatoire in Paris have undertaken measurements of the optical frequency of an optical or microwave cavity cooled to liquid helium temperatures. These experiments use the Earth itself to rotate the apparatus; if the (two-way) velocity of light were to depend on direction, the rotation would produce a shift in the resonance frequency, and relative shifts in cavities pointing in different directions can be

[42] The argument is taken from Brown and Pooley (2001). [43] Brillet and Hall (1979).

measured. The Humboldt group took data over a period of a year, and established a limit for variations in the light-speed of $\Delta c/c < 2 \times 10^{-15}$.[44]

A difficult and well-known variation of the MM experiment, first performed by Kennedy and Thorndike in 1932,[45] involved a Michelson interferometer with arms A, B of different lengths. Using a continuous, monochromatic source of light, the experiment was designed to detect over a six-month period any change in the brightness of the light emerging from a fixed, non-rotated, interferometer in the form of an alteration to the circular fringes produced in the eyepiece.

Suppose n is the (frame-independent) number of periods of the light waves associated with the time delay in traversing the two unequal arms. Relative to the laboratory frame S', this relative phase is

$$n = \Delta'/\tau' = 2(L'_A - L'_B)/c'\tau', \tag{5.8}$$

where τ' is the period of the light source, and Δ' is again (recall the discussion in subsection 4.2) the delay time, both relative to frame S'. Here, the length of arm A is assumed to be greater than the length of arm B. It is further assumed (in the light of the MM experiment) that the two-way light velocity c' relative to S' is isotropic. Notice that (5.8) holds even when the arms are not perpendicular (which was the case in the Kennedy–Thorndike experiment).

Any variation in the brightness of the emerging light measured over a significant period of time (i.e. one in which the Earth would be expected to significantly change its velocity relative to the ether) would indicate a variation of n with S', or with the velocity v. Since the conditions of the experiment are that L'_A, L'_B, and τ' are held fixed over the period in question (and maintaining this stability was the most difficult part of the experiment), any brightness variation must be a measure of the variation in c'. No appreciable variation was detected in the Kennedy–Thorndike (KT) experiment, in conformity with the Poincaré–Einstein relativity principle and the assumed constancy of the light velocity relative to S'. Thus, the inference from the combination of the MM- and KT-experiments is that the coordinate transformations take the form of the k-Lorentz transformations (4.26–4.29) for $|v| < c$. A repetition of the KT experiment using laser light was performed by Hils and Hall (1990); their null result held good to two parts in 10^{13}.[46]

Note that the length change factors (recalling the form (2.21–2.24) of the general linear inertial transformations) thus still take the forms imposed by the MM-experiment $\mathcal{C}_{\|} = (k\gamma)^{-1}$, $\mathcal{C}_{\perp} = k^{-1}$. But now the dilation factor takes the form $\mathcal{D} = \gamma/k$.

It is common to read in standard texts on SR that the KT experiment provides evidence for the relativistic dilation factor $\mathcal{D} = \gamma$, but this claim clearly only

[44] See Müller *et al.* (2003).
[45] Kennedy and Thorndike (1932).
[46] See Hils and Hall (1990).

holds if $k = 1$, and we have seen that this condition is not imposed by the MM experiment. (If the choice $k = \gamma$ is made, there is no dilation, at least relative to the S observer, and this is perfectly consistent with the KT experiment.) This fact was essentially recognized by Kennedy and Thorndike themselves, and stated quite explicitly in 1937 by Ives, and repeated by Robertson in 1949.[47]

5.6.2 The Situation So Far

Up until this point, we have fixed the deformation and dilation factors relative to the S frame, not the rest frame S' of the laboratory. What we really want are the corresponding factors defined relative to the latter frame.

We need to know the transformations between S' and S'', where S'' is a frame moving with arbitrary speed w along the positive x'-axis. Let us denote by V the velocity of S'' relative to S. Since the velocity transformation rule which follows from the k-Lorentz transformations does not depend on k (so it is the same rule as in SR) the velocities v, V, and w are related by

$$w = (V - v)/(1 - Vv/c^2). \tag{5.9}$$

In fact, since we know how to transform between S and S', and between S and S'' we can calculate the transformations from S' to S''. After some algebra we get

$$x'' = K\gamma(w)(x' - wt') \tag{5.10}$$

$$y'' = Ky' \tag{5.11}$$

$$z'' = Kz' \tag{5.12}$$

$$t'' = K\gamma(w)(t' - wx'/c^2). \tag{5.13}$$

where $K = k(V)/k(v)$ and $\gamma(w) = (1 - w^2/c^2)^{-1/2}$.

We are now able to state what the deformation and dilation effects are relative to S':

$$\mathcal{C}_{\|} = (K\gamma)^{-1}; \quad \mathcal{C}_{\perp} = K^{-1}; \quad \mathcal{D} = \gamma/K, \tag{5.14}$$

where $\gamma = \gamma(w)$. But we see that we still need to know not only what the functional form of k is, but also what the speed v is of the lab frame relative to the 'rest' frame S in the event that $k \neq 1$.

[47] See Ives (1937*a*) and Robertson (1949). Note that in their discussion of the KT experiment, Hsu and Zhang (2001, p. 527) incorrectly claim that 'the null result excludes the original hypothesis of the FitzGerald–Lorentz contraction of length'.

5.6.3 The 1938 Ives–Stilwell Experiment

Suppose the rest frequency is given for some source of monochromatic light, that is, the frequency relative to the source's rest frame. (We will assume, in accordance with the discussion in section 2.4, that the rest period does not depend on the state of motion of that frame.) If the source is at rest relative to the frame S'', so moving with speed w relative to the lab frame S', it is desirable to calculate the frequency of the light relative to the lab.

One might think that since for plane monochromatic waves, the frequency ν' is just the inverse of the period T' of the wave, then it is related to the rest frequency ν^0 in a way that is determined solely by the time dilation factor $\mathcal{D}$ in (5.14) and the direction of wave propagation. But this is not so. Recall first that for wave phenomena in classical kinematics there is a Doppler effect even in the absence of time dilation. Furthermore, the frequency of a plane wave is by definition measured by a single clock at rest in the relevant frame, whereas time dilation involves one clock at rest relative to one frame and two separated synchronized clocks at rest relative to another. Any transformation rule for frequencies concerns the readings of a single clock at rest in one frame and another at rest in another. The connection with time dilation is not obvious.

The transformation of frequency actually follows from the coordinate transformations together with the fact that the phase of a plane wave is invariant. (See the derivation of the relativistic Doppler effect in any relativity text.) The result for a plane wave moving in the x', y' plane say is:

$$\nu' = K\nu''/\gamma[1 - (w\cos\alpha)/c]. \tag{5.15}$$

where again $\gamma = \gamma(w)$, and α is the angle relative to the lab frame S' between the direction of the light emitted by the moving source and the x'-axis. (Note that although the source moves at speed w parallel to the x'-axis, the observer at rest in S' may choose to observe the light emitted from the moving source at any angle α.) But since in our example ν'' is the rest frequency ν^0 of the source, we have

$$\nu' = K\nu^0/\gamma[1 - (w\cos\alpha)/c]. \tag{5.16}$$

In 1938, H. E. Ives and G. R. Stilwell made measurements[48] related to the wavelength of light emitted by fast-moving ions for $\alpha = 0, \pi$, the rest wavelength of the light being known. Since wavelength is just the phase speed c divided by the frequency, the experiment tested the validity of (5.16) for these angles. The sensitivity of Ives and Stilwell's technique allowed for the second-order effect to

[48] Ives and Stilwell (1938).

be detected. The results were consistent with (5.16) for the value $K = 1$. Since it is highly likely that $V \neq v$, the implication is that $k = 1$.[49]

It is not unreasonable to conclude that it took until 1938 to establish experimentally the full form of the Lorentz transformations: that they form a group, that deformation is purely longitudinal ($\mathcal{C}_{\perp} = 1$) and that time dilation exists ($\mathcal{D} = \gamma$).[50]

But it would be quite wrong to give the impression that special relativity as a whole languished experimentally until the 1930s. An illuminating story surrounds the relativistic experiments performed in the period 1901–1905 by Walter Kaufmann (1871–1947). These involved measuring the deflection of fast electrons in strong magnetic fields. In the terminology of the day, Kaufmann was essentially measuring the dependence of mass on velocity—not a 'kinematical' effect. Kaufmann's results were not good for Einstein's SR, whose predictions coincided with Lorentz's, based on the notion of the deformable electron. The results were apparently closer to the predictions of Max Abraham (1875–1922), based on a theory of the electron which does not deform in motion. Lorentz was devastated, but characteristically faced defeat with true Popperian humility before the tribunal of experience. He wrote to Poincaré in March 1906:

> Unfortunately, my hypothesis of the flattening of electrons is in contradiction with the results of new experiments by Mr Kaufmann and I believe I am obliged to abandon it; I am at the end of my Latin and it seems to me impossible to establish a theory that demands a complete absence of the influence of translation on electromagnetic and optical phenomena.[51]

Einstein's response was much less defeatist. It was to cast doubt on the validity of Kaufmann's results—not only were the competing theories less plausible than his, or so he argued, but he had a shrewd idea about possible systematic errors in Kaufmann's procedure. Einstein's defiance was vindicated when new versions of the experiment performed by Bucherer in 1909, and particularly by Neumann in 1914 and by Guye and Lavanchy in 1915, produced results consistent with the relativistic predictions.[52]

Karl Popper's famous criterion of demarcation between science and pseudoscience, based on the empirical falsifiability of theories, was partly inspired by Einstein. Popper was greatly impressed by Einstein's bold conjecture in his general

[49] Recent more accurate versions of the Ives–Stilwell experiment were performed by Hasselkamp *et al.* (1979) and Kaivola *et al.* (1985). A review of the three classic experiments discussed in this section is found in Hsu and Zhang (2001), Part II.

[50] For a recent review of Lorentz covariance tests, see Pospelov and Romalis (2004). For new tests motivated by quantum gravity considerations, see Amelino-Camilia and Lämmerzahl (2004).

[51] Cited in Galison (2004), p. 220.

[52] For nice treatments of the whole episode, see Cushing (1981), Hon (1995), and Staley (1998). It is worth noting that in the opinion of L. Jánossy, it was not before 1940 that experimental results were accurate enough to decide between the Abraham predictions and those of Lorentz and Einstein; see Jánossy (1971), p. 45, and further references therein.

theory of relativity regarding the redshift of spectral lines of atoms in a gravitational field.[53] Popper contrasted Einstein's willingness in 1917 to stick his neck out, and face the possibility of empirical refutation, with the behaviour he regarded typical of social scientists and particularly psychologists and psychoanalysts. At any rate, it has been claimed that 'apart from Einstein, Popper probably did more than any other individual to change the twentieth century's conception of science.'[54] It is hard not to wonder how the history of twentieth century philosophy of science might have gone if the young Popper had known about Einstein's reaction to the Kaufmann experiments.[55]

5.7 ARE EINSTEIN'S INERTIAL FRAMES THE SAME AS NEWTON'S?

It is a remarkable fact that the experimental results discussed so far which determine the nature of the coordinate transformations between inertial frames do not pin down precisely what the inertial frames are. In particular, they do not help us decide whether a frame co-moving with the Earth's surface (for a sufficiently short period of time) or with a freely falling particle is a better inertial frame. It is the so-called *terrestrial redshift experiments* that provide the answer.[56]

In the famous redshift experiment of R. W. Pound and G. A. Rebka reported in 1960, at the top of a 74-foot tower an accurate measurement was made (using the Mössbauer effect) of the frequency of γ-ray photons emitted from the base of the tower. This result was compared with the frequency of similar photons emitted by the same source (^{57}Fe nuclei) moved to the top of the tower. Now if the tower is at rest relative to an inertial frame, Maxwellian electrodynamics predicts that there will be no shift in frequency, and hence wavelength of the γ-rays. However, if the tower is accelerating uniformly a frequency shift is predicted using the expression for the transverse Doppler shift in SR. (The Doppler effect is caused by the motion of the absorber relative to the emitter, at the time of absorption, due to its acceleration during the time of transmission.) In particular, if inertial frames

[53] See Popper (1982), p. 38.

[54] Bryan Magee, in the *Financial Times*, 19 September, 1994.

[55] Cushing's summary (op. cit., p. 1133) of the Kaufmann story reads as follows.

The entire episode provides another example that science does not proceed by a strict falsificationist methodology. It shows rather that that a great scientist such as Einstein at times gives more weight to a theory that has a certain beauty and produces equations simple in form than he does to experimental results that apparently conflict with such a theory.

[56] This section owes much to Ehlers (1973), §1.4. For useful analyses of the observations made by a uniformly accelerated observer in SR, see Hamilton (1978), Giannoni and Gron (1979), and Desloge and Philpott (1987).

are attached to freely falling particles, then the predicted redshift in wavelength is

$$\frac{\Delta\lambda}{\lambda} = \frac{gl}{c^2} + \text{much smaller terms,} \tag{5.17}$$

where g is the free-fall acceleration relative to the surface of the Earth and l the height of the tower. The results of Pound and Rebka were in good agreement with this shift; a more precise version of the experiment by Pound and Snider confirmed it in 1964 with an accuracy of 1 per cent.[57]

There are several aspects of this result that are worthy of emphasis. The first is that the nuclei in the experiment, which act both as emitters and absorbers of the γ-ray photons, are assumed to act as ideal clocks in spite of the fact that they are accelerating relative to the free-fall, inertial frames. Nothing so far in Einstein's SR refers to the behaviour of accelerating rods and clocks, so clearly some kind of extra ingredient is needed if the theory is to be applicable to this case. This is the so-called *clock hypothesis*, whose significance and justification will be discussed in the next chapter. (It is an essential element of any decent discussion within SR of the so-called Twins Paradox, or the clock retardation effect.) The important thing to note at this stage is that it does not require for its justification any appeal to general relativity.

The second point is that the present reading of the significance of the redshift experiments presupposes that an inertial frame is one relative to which Maxwellian electrodynamics are valid. (I do not mean this in an exact sense; what is really being assumed is that the standard laws of quantum electrodynamics are valid relative to the freely falling frame; it is just that Maxwell's equations are a perfectly good approximation in this experiment.) It could hardly be otherwise, given the origins of SR. It is the behaviour of matter (including fields) that determines inertial structure in SR, not the other way round. But if, in the spirit of the relativity principle, we make the further highly non-trivial assumption that the fundamental equations of all the other non-gravitational interactions also pick out the freely falling frames in this way, then a profound implication emerges. Since a freely falling particle does not, it turns out, accelerate relative to the inertial frames in SR, gravity cannot be a force in the usual Newtonian sense. Of course, Newtonian gravity was always going to be a problem for SR, given that it is an instantaneous action-at-a-distance and hence not Lorentz covariant. But now it seems gravity is not even an action! It further follows that Einstein was wrong in the 1905 paper to identify his inertial frames with Newton's. So his RP turns out not to be equivalent to Newton's after all. They sound pretty much exactly the same, as we saw in section 5.3.1. But the terrestrial redshift experiments show that they refer to different families of frames.

Finally, there is the obvious, and now rather alarming, fact that free particles falling at different places on the Earth's surface are in relative acceleration, the

[57] For a fuller discussion see, e.g., Ohanian and Ruffini (1994).

magnitude of the acceleration depending of course on the distance between these places. Inertial frames are thus either impossible, or they must be curtailed in the sense that they are considered well-defined only 'locally'. We are now so steeped in general relativity that the notion of local inertial ('geodesic') frames seems natural and straightforward. We often forget how tricky this notion of 'local' really is. But what is of greater interest at this stage is the question as to how the replacement of global frames by a multitude of local ones is supposed to work.

The mathematical theory of Riemannian spaces incorporates the notion of an affine connection which allows comparison to be made, generally in a path dependent way, between tangent vectors at the different locations in the space. Perhaps space-time comes equipped with such an 'affinity' that allows for the inertial structure at different events to be systematically connected. If the affine connection is dynamical, subject to field equations which couple it to matter fields, then a theory of what is commonly called the gravitational force but now understood as space-time curvature may emerge which explains the stupendous success of Newton's theory despite its misconception about the nature of gravity. We cannot be sure at this stage how the dynamics of the pure 'gravitational' degrees of freedom will incorporate this and possibly other affinities. But if the programme as it has been broadly defined is successful, as Einstein showed it was in 1915,[58] one question deserves particular attention. Does the seeming miracle of all the non-gravitational forces concurring about local inertial structure receive an explanation in this programme?

5.8 FINAL REMARKS

Let's return finally to Einstein's 1949 *Autobiographical Notes* and the oddity flagged at the end of section 5.1 above. The whole point of the principle theory approach to relativistic dynamics that Einstein consciously adopted in 1905 was to find phenomenological principles, akin to the laws of thermodynamics, which would constrain the behaviour of rods and clocks. Now imagine in thermodynamics if the second law were replaced by the assertion that there exists a state parameter called entropy, which over time would in the case of adiabatic processes either remain constant or increase. Normally of course, this is taken to be a theorem of the standard laws, but now we consider postulating it. It is not a phenomenological law; it cannot be directly distilled from a large amount of empirical experience, like the impossibility of constructing a perpetual motion machine of the second kind. It has moved in the direction of the abstract. The new version of the theory shifts along the theory continuum that has classical thermodynamics at one extremity and statistical mechanics at the other. It is still far from supplying the explicit

[58] Actually, in 1915 Einstein was not aware of the significance of the affine connection in his theory of gravity, as we see in chap. 9.

micro-structure that statistical mechanics, or even the kinetic theory of gases, provides. But it represents a displacement away from the pure principle theory that is instantiated in classical thermodynamics.

Something like this shift seems to occur in the *Autobiographical Notes* when, a few pages after announcing the need to find a universal principle analogous to the laws of thermodynamics, Einstein writes:

> The universal principle of the special theory of relativity is contained in the postulate: The laws of physics are invariant with respect to the Lorentz-transformations (for the transition from one inertial system to any other arbitrarily chosen system of inertia). This is a restricting principle for natural laws, comparable to the restricting principle of the non-existence of the *perpetuum mobile* which underlies thermodynamics.[59]

But the comparison with thermodynamics has now been weakened. I think that there can be no doubt that initially Einstein thought of the RP and the LP, or their combination, as the analogues of the laws of thermodynamics—recall the discussion in section 5.2 above. In 1905, and to a lesser but still significant extent in 1949, when Einstein wrote the above, the LP was based on relatively little *direct* empirical evidence, certainly less than the RP. But both principles are phenomenological in character, as was repeatedly stressed above. In particular, the LP does not depend on whether light is made of wave, particle, or a combination of the two; like the RP it is taken to be independent of the micro-structure of matter and radiation. The requirement of the Lorentz-covariance of 'natural laws', of the equations that govern the fundamental interactions of nature, is certainly not phenomenological. True, it does not make a commitment to the *exact* form of the deep structure of matter and radiation; it is, as Einstein says, a restricting principle on any theory purporting to describe this structure. But the requirement was also explicitly defended by Poincaré, and clearly represents a shift away from the pure principle theory concept that was unique to Einstein. The stipulation of Lorentz-covariance was not, nor could it have been, the starting point of Einstein's 1905 distinctive approach to SR.

What, then, is special relativity? This question is taken up in Chapter 8, after we have dealt with some preliminary matters.

[59] Einstein (1969), p. 57.

6

Variations on the Einstein Theme

Had silicon been a gas, I would have been a major general.

James McNeill Whistler[1]

6.1 EINSTEIN'S OPERATIONALISM: TOO MUCH AND TOO LITTLE?

The operational nature of Einstein's discussion of inertial coordinate systems, and the issue of distant simultaneity in particular, has been hugely influential. In physics, it was consciously emulated by the founder of the matrix mechanics approach to quantum mechanics; recall Heisenberg's dismay in finding that Einstein of all people did not embrace his operationalist stance in the late 1920s. In philosophy, it was influential in the emergence of the doctrine of operationalism and played a significant role in the development of logical positivism. But after the pendulum swung back and operationalism (or rather its radical version due to such writers as Percy Bridgeman) got such a bad name in philosophy, it has been fashionable for some time in the philosophical literature to discuss space-time structure without any reference at all to such base elements as rods and clocks. This has been unfortunate. It has tended to prevent philosophers from asking the important question as to why real rods and clocks happen to survey the postulated space-time structure.[2] All in all, Einstein's treatment has at times been taken too literally by both physicists and philosophers.

The point was made earlier that a typical physics laboratory contains no grid-like arrangement of rigidly connected synchronized clocks, not to mention no collection of rigid rulers hanging around to determine precisely where everything is. (Just think of the early experiments related to SR, such as the MM and KT experiments.) This point is banal.[3] What Einstein presumably had in mind was

[1] The artist made this remark after he failed his studies at the West Point military academy in 1854 because of his ignorance of chemistry.

[2] We shall return to this issue in the next chapter.

[3] It seems a little less banal, though, when it is borne in mind that if a laboratory representing an inertial frame really did have a system of rigidly connected synchronized clocks, and the lab was boosted to a new state of uniform motion, then the clocks would cease to be Einstein synchronized as a result of the joint acceleration; see Giannoni and Gron (1979).

a counterfactual claim: the temporal coordinates associated with distant events relative to some inertial frame *would* be read off by suitably synchronized clocks *were they there*, and similarly with positions and rulers. But is clear, as we saw in section 5.5, that Einstein considered the rods and clocks as *defining* distances and temporal intervals, and it is this notion that is highly questionable. In particular, it makes proper time seem more fundamental than (inertial) coordinate time, when the opposite is the case.

But if we appeal, as Einstein did, to hypothetical, ideal clocks and rods, it is beholden on us to say what they are, and here, to repeat, Einstein was strangely silent, at least in 1905. In 1910, he was more forthcoming. 'By a clock we understand anything characterized by a phenomenon passing periodically through identical phases so that we must assume, by virtue of the principle of sufficient reason, that all that happens in a given period is identical with all that happens in an arbitrary period'.[4]

The problem, of course, is to spell out how the principle of sufficient reason should be applied systematically and not just on the basis of whim. If one tries to fill in the blanks in Einstein's definition, one is inevitably drawn into dynamics, into a discussion of the simplest form of the dynamical equations governing the fundamental processes under investigation. Spatial and temporal coordinates are chosen on the basis of such simplicity considerations, and rods and clocks are then defined to be bodies that best take up space and tick, respectively, in accordance with these natural coordinates. Good conceptual housekeeping means that coordinatization comes before operationalism in Einstein's sense, and it seems that Poincaré had a better understanding of this point than Einstein.[5]

6.2 WHAT IS A CLOCK?

George F. FitzGerald was also one to stress the dynamical nature of time. In an essay written before the turn of the century on the nature of physical measurement, and never published in his lifetime,[6] FitzGerald gave a lucid account of the problem of defining temporal duration in physics. The standard of time universally adopted in FitzGerald's day was the time of rotation of the Earth, and he pointed out that although 'there is every reason for assuming that the Earth rotates on its axis more uniformly than any clock we can construct', there is nonetheless good ground for thinking that this rotation rates is slowing down.[7] Not only do the

[4] Cited in Galison (2004), p. 266.

[5] Recall the discussion of Newtonian time in section 2.2.4.

[6] See FitzGerald (1902); the essay was published in Larmor's edition of FitzGerald's collected papers, with the editorial comment 'Hitherto unpublished: apparently the introduction to a projected treatise on physical measurement.' The mention on p. 536 of advances in the science of heat 'during the present century' is a clear indication that it was written before 1900.

[7] Actually, although the long term trend has been a slowing down, over the last decade the rotation has been speeding up. See Bradt (2004), p. 86.

records of ancient eclipses testify to this, said FitzGerald, but it is plausible that frictional resistance to the tides caused by the sun and moon are responsible for the slowdown. But this situation raises a conceptual conundrum: '. . . how on earth can we discover a change in our standard itself?'[8]

The answer of course is by finding a 'more ultimate standard of time' postulated in the investigation of the Earth's rotation. This is none other than that defined by the uniform rectlinear motion and rotation of free bodies in Newtonian mechanics, and the assumption that Newton's law of gravitation accurately holds for the solar system. 'Hence our real ultimate standard of time is an assumed accuracy of these laws, and there is a great deal to be said in favour of explicitly basing our measure of time upon these laws of motion and gravitation. This is possible and has been more than once suggested.'[9]

FitzGerald was anticipating what came to be called *ephemeris time* (originally *Newtonian time*) in astronomy. By the late 1930s, discrepancies in the observed motion of the Sun, Moon, and planets led to the abandonment of the standard of time based on the Earth's rotation.[10] By the 1950s astronomers had officially adopted the standard defined by ephemeris time: the independent variable in the equations of motion of the bodies of the solar system. By monitoring one body, the Moon, they could check when it reached positions predicted on the basis of these laws and verify that the other bodies in the solar system reached their predicted positions. They used, in Julian Barbour's words, 'the Moon as the hand of a clock formed by the solar system'.[11] Ideal clocks were then defined as those which marched in step with this ephemeris time.

But there are practical limitations to the accuracy with which ephemeris time can be measured, and by 1967 atomic clocks were officially adopted to provide the fundamental unit of measurement of the second.[12] An important feature of

[8] op. cit., p. 538. [9] ibid.

[10] The rotational motion of the Earth has five components: uniform motion, forced secular motion (precession) that was recognized by Hipparchus, Euler motions (the Chandler wobble), periodic motions (nutation), and finally irregular motions that were suspected from the eighteenth century onwards. (See Kinoshita and Sasao (1989).) The last of these, due to geophysical phenomena on the surface and within the earth, are unpredictable. FitzGerald mentioned frictional resistance to the tides as the cause, as we have seen, but the true situation is much more complicated.

> A complete understanding of the driving mechanisms requires a study of the deformation of the solid Earth, of fluid motions in the core and of the magnetic field, of the mass redistributions and motions within the oceans and atmosphere, and of the interactions between the solid and fluid regions of the planet. (Lambeck (1989).)

[11] Barbour (1999), p. 107. For further details see Taff (1985), pp. 100–2. By the 1980s, lunar ranging by lasers was accurate enough to require general relativistic corrections to Newtonian predictions for the motion of the moon.

[12] Useful accounts of the different measures of time used in astronomy are given in Sadler (1968) and Bradt (2004), section 5.4. They remind us that a clock used to establish the fundamental unit of measurement of the second is not the same as a procedure for determining duration.

> If I wanted to know how many . . . seconds (atomic time) I had lived, I would find the number of days since my date of birth, taking into account leap years . . . . I would adjust this for partial days at each end of the interval and then multiply by 864400 s/d. Finally I would add the difference

the use of atomic clocks must be stressed. A single high-quality caesium clock is not enough, because it is recognized that there is a risk of failure or abnormal functioning. In establishing an international atomic time scale, an *averaging over a number of such clocks in different laboratories* is used, with appropriate statistical weighting for the 'best' clocks. Moreover, stability algorithms for atomic time scales are employed in each laboratory.[13] Such procedures highlight the fact that *it is theory that rules in the construction of accurate clocks, not the availability of accurate clocks that rules in the construction of theory.*[14] And the same holds for rigid rods.

6.2.1 The Clock Hypothesis

So far we have been discussing clocks that move inertially, or equivalently, that are isolated from their environment. What happens when they accelerate in relation to inertial frames? If the clock is sufficiently small, it defines a world-line or track in space-time which is 'straight' if left to itself, but crooked if accelerating and hence being subject to external forces. Eddington put his finger on the problem in relation to the latter case: 'We may force it [the clock] into the track by continually hitting it, but that may not be good for its time-keeping qualities.'[15]

An important part of the history of time has been the search for accurate clocks which withstand buffeting, which, in short, can be carried around. In particular, in the field of maritime navigation, the ability to measure longitude at sea was severely handicapped, often with tragic consequences, for centuries by the lack of accurate clocks that could withstand the tossings of the seas.[16]

An atomic clock can in principle be adversely affected by tidal gravitational forces acting on either the atomic process responsible for the frequency stability or on the locking to the the crystal oscillator responsible for the periodic signal output, the latter being disrupted long before the former.[17] But such forces are negligible in the solar system, and of more practical interest is the effect of

> in the two TT–UTC offsets at the ends of the interval. . . . The latter step takes into account the leap seconds that were inserted between the two dates. (Bradt (2004), p. 87)

Here TT stands for Terrestrial Time (ephemeris time corrected for atomic time) and and UTC stands for Universal Coordinated Time (time based on the earth's rotation corrected for atomic time). For more detailed accounts of time in astronomy and the use therein of atomic clocks, see Guinot (1989*a*, *b*). It is interesting that some rotating neutron stars, seen as regularly pulsing *radio pulsars*, have an extremely high degree of stability, rivalling that of atomic clocks; see Bradt (2004), p. 91.

[13] See Guinot (1989*a*), p. 390. For a popular account of the 'best' clock in the world in 2000, see Klinkenborg (2000).

[14] There is a widespread view, possibly due originally to Erwin Schrödinger, that no clock modelled in quantum mechanics can be perfect even in principle. This is because any such clock must possess an energy spectrum that ranges (not necessarily continuously) from minus to plus infinity. (See in this connection Unruh and Wald (1989), pp. 2605–6.) A recent careful analysis of the matter is given in Hilgevoord (2005), which contests the standard view and plausibly concludes that ideal clocks are indeed possible in quantum mechanics.

[15] Eddington (1966). [16] See Sobel (1996). [17] See Misner *et al.* (1973), p. 396.

acceleration. It is important to appreciate that this effect will vary depending on the nature of the clock, even in the case of atomic clocks. The key issue is the comparison of the magnitude of the external force producing the acceleration and that of the forces at work in the internal mechanism of the clock. Even moderate accelerations can disturb the locking mechanism in many atomic clocks. But other kinds of clocks, such as those based on certain nuclear frequencies, can continue to function 'ideally' under accelerations of up to 10^8 cm/s^2.[18]

If the accelerative forces are small in relation to the internal restorative forces of the clock, then the clock's proper time will be proportional to the Minkowski distance along its world-line. Consider two events A and B lying on this time-like world-line. The distance along the world-line between these events is given by $\int_A^B ds$, where $ds^2 = c^2dt^2 - dx^2 - dy^2 - dz^2$ in inertial coordinates. It is a sum, in other words, of 'straight' infinitesimal elements ds: the effect of motion on the clock depends accumulatively only on it instantaneous speed, not its acceleration. This condition is often referred to as the *clock hypothesis*, and its justification, as we have seen, rests on accelerative forces being small in the appropriate sense. The term 'hypothesis' is arguably a misnomer, and tends to hide the straightforward dynamical issues at stake. There should be no mystery as to why clocks are way-wisers of space-time.[19]

Indeed, it is noteworthy that in many accounts of the behaviour of a clock in motion, dynamical considerations seem relevant, if at all, *only* when the clock is accelerating. To be sure, in the case of inertial motion, the behaviour of the clock is universal, and does not depend on its constitution, whereas whether a given force acting on a clock renders it unreliable depends crucially on the make-up of the time-piece. But if it is argued that it is the nature of space-time itself that accounts both for the stability (and time dilation) of the clock in inertial motion, when it comes to accelerative motion it seems to be a competition between the 'pull' of space-time and the effect of the applied force. How are they to be compared? Like is not being compared with like.

6.3 THE CONVENTIONALITY OF DISTANT SIMULTANEITY

It is a remarkable thing that distant simultaneity has proved over the years to be one of the most contentious and apparently confusing issues in the conceptual

[18] See Ohanian and Ruffini (1994), p. 172.

[19] There is a strict analogy with the notion of distance in Euclidean three-dimensional space. Consider a curve joining two points A and B, whose distance is taken to be $\int_A^B ds$, where $ds^2 = dx^2 + dy^2 + dz^2$ in Cartesian coordinates. Why does a string placed along the curve measure this distance? Because, to put it crudely, it has the microscopic structure of a bicycle chain: at the atomic level the size of smallest links does not depend on the degree of bending of the string. (This point became clear to me in discussions with Jeeva Anandan.)

foundations of relativity theory. It is true that it was the last piece of the jigsaw that Einstein hit on in formulating SR in 1905, and in particular in showing that his two postulates are not mutually exclusive. But does the issue deserve to be such so controversial?

Let us start by examining non-standard definitions of simultaneity. Recall the transformation (2.6) in section 2.2.3 involving a change in the simultaneity relation for a given inertial frame S:

$$\tilde{\mathbf{x}} = \mathbf{x}; \quad \tilde{t} = t - \vec{\kappa}.\mathbf{x}. \tag{6.1}$$

for any constant 3-vector field $\vec{\kappa}$. It was pointed out that this transformation is a symmetry of the equation of motion for the free particle. But there is a broader issue to be addressed. Suppose $\vec{\kappa}$ is no longer constant, but depends on space. Suppose furthermore that clocks assigned *à la* Einstein to points in space are resynchronized in terms of $\vec{\kappa}$. Would any observable differences occur in any dynamical theory submitted to such a reformulation? Not as long as the equations of motion of the relevant dynamical entities undergo a corresponding $\vec{\kappa}$-dependent reformulation, and this certainly takes place if the theory can be formulated in the tensor calculus.[20]

In the case of special relativity, consider the effect of the transformation (6.1) when the coordinate t is the usual one resulting from the Poincaré-Einstein convention for synchronizing clocks in frame S. After the transformation, the velocity of light in the direction $\hat{\mathbf{n}}$ is now

$$\tilde{c}(\hat{\mathbf{n}}) = \frac{c\hat{\mathbf{n}}}{1 - c\vec{\kappa}.\hat{\mathbf{n}}}. \tag{6.2}$$

Now in many discussions, the space-time is taken to be $(1+1)$-dimensional and the vector field $\vec{\kappa}$ is taken to be constant, resulting in

$$\tilde{c}_{\pm} = \frac{c}{1 \mp c\kappa}. \tag{6.3}$$

In the philosophy literature, it is common to use a different parametrization of synchrony (or better, the anisotropy of the light speed) due to Hans Reichenbach based on the parameter ϵ, where

$$\epsilon = \frac{1}{2}(1 + c\kappa), \tag{6.4}$$

[20] An illuminating discussion of this case is found in Anderson *et al.* (1998), in which it is shown how the components of the 4-metric are altered by the components of $\vec{\chi} \equiv \nabla(\vec{\kappa}.\mathbf{x})$ and how the irrotational (curl-free) nature of $\vec{\chi}$ ensures the synchrony independence of the round-trip speed of light; see section 2.3.4.

so that $\tilde{c}_+ = c/2\epsilon$ and $\tilde{c}_- = c/2(1-\epsilon)$. The advantage of this parametrization is easy to see. Suppose event A represents the emission of a photon, B its reflection at a distant mirror, and C its absorption back at the same (spatial) point from where it was emitted. There is no question as to how long the round-trip journey took, call it T; it can be measured by a single inertial clock whose world-line contains the events A and C. The question is: relative to the S frame, when did the event B take place? If A occurs at time t_A and thus C at $t_C = t_A + T$, then $t_B = t_A + \epsilon T$. Poincaré–Einstein synchronization is clearly given by $\epsilon = 1/2$.

I will have more to say about this Reichenbach factor ϵ shortly, but note that it is widely assumed that ϵ must be restricted to the closed set [0, 1], or equivalently $|c\kappa| \leq 1$. This is to ensure that in one direction light does not propagate backwards in time. It is often claimed that such a possibility would violate the fundamental canons of causality, but it is a hum-drum experience for airline travellers flying East across the International Date Line.[21] No, such a restriction on the parameters is again mere convention. But recognizing this point in the case of special relativity severely weakens the difference between SR and Newtonian mechanics vis-à-vis the conventionality issue. We saw in section 2.2.3. that any liberalization of the standard convention in Newtonian mechanics for spreading time through space would result in the gravitational action-at-a-distance no longer being instantaneous, even though a round-trip gravitational effect would be. The idea of a gravitational effect occurring before its cause in this sense is no better or worse than the idea in SR of a photon arriving at a detector before it left its source. We standardly choose temporal coordinates in such a way as to prevent this odd-sounding possibility (unless we are forced to spread time non-uniformly across a finite two-dimensional surface with the global topology of a sphere, like the surface of the Earth) not because it is metaphysically imperative, or empirically established, but because it is convenient.[22] The only remaining difference between Newtonian mechanics and SR is that in the latter one is still left with a degree of conventionality: the choice of κ consistent with $|c\kappa| \leq 1$, or of ϵ within the interval [0, 1], and again the criterion is convenience. As has been pointed out on a number of occasions above, in the light of the MM experiment and its modern variants which establish the isotropy of the two-way light speed, it is natural to adopt the convention according to which the one-way speed is also isotropic.

There is still confusion about this last issue. Perhaps the most significant instance is associated with attempts to define 'test theories' of SR, such as the 1977 effort due to Mansouri and Sexl. In this approach, although it is accepted

[21] This nice point is made by Anderson *et al.* (1998), sections 1.5.1 and 2.3.2. I can testify, having flown from New Zealand to both North and South America, that arriving before you left is survivable! (Note that the establishment of international time zones is a case of a spatially dependent and discontinuous κ factor.) Come to think of it, every telephone call from, say Australasia to the UK, involves a signal arriving before it left, and no one seems the worse for it.

[22] The odd kind of possibility being discussed here has roughly the status of those singularities in general relativity which can be transformed away under a suitable coordinate transformation.

that the one-way speed of light is not measurable in SR, it becomes measurable in other elements of the theory-continuum defined by values of test parameters associated with space-time structure. This line of thinking has been widely endorsed. However, a careful analysis of such test theories, and many other related issues, has been given in a lengthy 1998 review paper by Anderson, Vetharaniam, and Stedman, in which the case for the conventionality of distant simultaneity is in my opinion made overwhelmingly.[23] But not even this *tour de force* has succeeded in settling the issue.[24]

6.3.1 Malament's 1977 Result

I cannot discuss the issue of distant simultaneity without devoting some words to a result which virtually single-handedly managed to swing the orthodoxy within the philosophy literature from conventionalism to anticonventionalism.[25]

David Malament's 1977 result was an extension of much earlier work of Alfred Robb, an English mathematician interested in axiomatizing Minkowski space-time using just the relation of causal connectibility between space-time points—in other words, using the light-cone or conformal structure of space-time.[26] Robb was able to define a 4-dimensional notion of orthogonality which, it turned out, had a clear link to the Poincaré–Einstein convention. Specifically, imagine an inertial world-line W and any point p on W; then the set of points q such that the straight line joining p and q is orthogonal to W in Robb's sense turns out to be just the set of all points simultaneous with p according to the Poincaré–Einstein convention in the inertial rest frame of the free particle whose world-line is W. (It is being assumed that light would propagate along light-like curves, even though no geometrical object representing the electromagnetic field appears in the argument. Note too that the notion of straightness that applies to both W and the curve joining p and q is based on the affine structure that is definable in terms of the postulated conformal structure.[27]) Malament's famous result is that 'Robb-orthogonality'[28] is the only non-trivial way of defining simultaneity solely in terms of W and the causal structure of space-time.

[23] See Anderson *et al.* (1998).

[24] A very recent attempt to defend the anti-conventionalist position is found in Ohanian (2004). The essence of the argument rests on the erroneous claim that choosing a non-standard synchrony convention should give rise to measurable effects analogous to the centrifugal effect associated with rotating coordinate systems.

[25] Norton (1992) described the impact of the result as 'one of the most dramatic reversals in the debates in the philosophy of space and time'.

[26] See Malament (1977) and Robb (1914). The following characterization of Malament's result and its connection with Robb's work follows Rynasiewicz (2001). See also in this connection Lucas and Hodgson (1990), §3.3, and Anderson *et al.* (1998).

[27] See, e.g., Friedman (1983), p. 164. It has been pointed out to me by Rob Rynasiewicz that whether the affine structure is definable in terms of the conformal structure depends on the global properties of the manifold; in the case of R^4 it is definable.

[28] This terminology is due to Rynasiewicz (2001).

Malament's 1977 paper is a model of logical and mathematical rigour, and it detracts little from the theorem to say that the result is unsurprising. Consider any two distinct points on the world-line W, and imagine the half light cones emanating from each point that contains the other point. The intersection of these cones defines a flat 'space-like' hypersurface all of whose points are Robb-orthogonal to W, and it is hard to see how any other such non-trivial structure could emerge from the very limited means available.[29]

Malament's stated objective was to provide a clean refutation of an earlier, specific claim by Adolf Grunbaum that simultaneity relative to an inertial observer is not uniquely definable in terms of the relation of causal connectability. But within the philosophical community, Malament's result has often, perhaps mostly, been given a wider significance; essentially because the conformal structure is in turn definable in terms of the metric structure of Minkowski space-time, the result has commonly been taken to settle the conventionality issue once and for all.[30] In his *Foundations of Space-Time Theories*, Michael Friedman was clear as to the ramifications of the result.

> So we cannot dispense with standard simultaneity without dispensing with the entire conformal structure of Minkowski space-time. Second, it is clear that if we wish to employ a nonstandard [simultaneity] ... we must add further structure to Minkowski space-time. ... This additional structure has no explanatory power, however, and no useful purpose is served by introducing it into Minkowski space-time. Hence the methodological principle of parsimony favors the choice of Minkowski space-time, with its 'built-in' standard simultaneity, over Minkowski space-time plus any additional nonstandard synchrony.
>
> These considerations seem to me to undercut decisively the claim that the relation of [simultaneity] ...is arbitrary or conventional in the context of special relativity.[31]

In a similar vein, Roberto Torretti, in his monumental *Relativity and Geometry*, expressed the lessons of the Malament result as follows:

> In the natural philosophy of Relativity, time relations between events are subordinate to and must be abstracted from their space-time relations. A partition of the universe into simultaneity classes amounts to a decomposition of space time into disjoint space-like hypersurfaces each one of which separates its complement into two disconnected components. ... Such a decomposition is impossible unless the universe is stably causal; but where one is possible many more are available as well. Not all of them, however, will

[29] This way of putting things makes it clear that Malament requires that the simultaneity requirement be invariant under temporal reflection, and this requirement has been questioned by Sarkar and Stachel (1999). Actually, Sarkar and Stachel argue that non-trivial simultaneity relations other than that associated with Robb-orthogonality are definable in terms of W and the given causal structure, so perhaps we should not regard Malament's result (as opposed to its proof) as entirely obvious. Indeed, there are subtleties concerning the notion of 'definability' in Malament's argument, as is shown in the careful criticism of Sarkar and Stachel's claims given by Rynasiewicz (2000*b*).

[30] Not all philosophers of physics, however, have taken this line; see for instance Debs and Redhead (1996), Anderson *et al.* (1998), Sarkar and Stachel (1999), and Rynasiewicz (2001).

[31] Friedman (1983), p. 312. See also Dieks (1984) for a similar claim in relation to the Malament theorem.

be equally significant from a physical point of view. ...In the Minkowski space-time of Special Relativity, the hypersurfaces orthogonal to the congruence of timelike geodesics which makes up any given inertial frame define a partition that is naturally adapted to that frame.[32]

It is important at this point to emphasize that both Friedman and Torretti are fully aware that the adoption of a temporal coordinatization of Minkowski space-time that fails to respect the simultaneity hyperplanes picked out by Robb-orthogonality in relation to the world-line *W* will not lead to anything so drastic as predictive error. Friedman in particular stresses the point that 'Minkowski space-time can be described equally well from the point of view of any coordinate system': the empirical equivalence of standard and non-standard coordinatizations 'reveals no deep facts about Minkowski space-time and special relativity; rather it is simply a trivial consequence of general covariance'.[33] Indeed, anti-conventionalists almost always concede this point, but take such justification of the possibility of coordinates not adapted to Robb-othogonality to be 'trivial'. Friedman insists that what is important is not mere description, but the nature of the coordinate-independent geometric structures, and how they illuminate the issue of simultaneity.

I do not find the logic of this argument compelling, insofar as I understand it. Largely this is because I see the absolute geometrical structures of Minkowski space-time as parasitic on the relativistic properties of the dynamical matter fields, a view which will be articulated in the following chapter, and which makes the 'non-trivial' take on the issue of simultaneity given by Friedman and Torretti appear to be question-begging. The essential point can be made, however, in the following relatively simple way.

Why should we consider defining simultaneity just in terms of the limited structures at hand in the Grunbaum–Malament construction, namely an inertial world-line *W* and the causal, or light-cone structure of Minkowski space-time?[34] Part of the answer is already obvious in Malament's paper: *W* is taken to represent an inertial observer, and we are after all talking about simultaneity relative to such an observer. But in the real world there is a lot more structure for the observer to observe: is none of this relevant? To help clarify this issue, let's start with what I shall call the Malament world, consisting of literally nothing more than *W* and the causal structure discussed in his paper.

The Malament world is so utterly different from ours, I think it is legitimate to ask whether it even contains time at all.[35] It is not enough to say that being

[32] Torretti (1983), p. 230. [33] Friedman (1983), p. 175.

[34] The same question has been posed by Rynasiewicz (2001), but his justification seems to be quite different from mine.

[35] In assessing the significance of Malament's result, I am reminded of a saying I heard from my old teacher, the late Heinz Post, who was a sceptic about the meaningfulness of wildy counterfactual claims: 'If my grandmother had four wheels, she would be a bus.' Another delightful counterfactual, for which I thank Ronald Anderson, S. J., is in the epigraph at the beginning of this chapter.

four-dimensional, the space-time manifold therein has time built into it. We are doing physics, not mathematics. (A similar point was made in Chapter 2 in relation to the 4-dimensional projective space used by some commentators to articulate the meaning of Newton's first law. Note that a similar concern can be expressed in relation to the conformal structure of the manifold in the Malament world. It is a misnomer of breathtaking proportions to use the adjective 'causal' in a world where neither action nor reaction takes place.) Time, at its most fundamental level, has something to do with change, and change is not an obvious feature of the Malament world. There are two absolute, non-interacting (but to some degree correlated) structures. The conformal light-cone structure is in itself timeless. It has no non-trivial dynamics. Supposedly there is also a particle or observer in motion, but in motion relative to what? There can only be one answer: in relation to the space-time manifold. But if Malament's world is anything at all like ours, this is not a notion that today, after the lesson of Einstein's hole argument has finally sunk in, is widely regarded as physically meaningful. Putting the point another way, take any two instantaneous configurations or states of the universe associated with different points along the particle world-line using Robb-orthogonality (or any other procedure for that matter): it is unclear whether they are discernible and whether they should not be identified physically.

I think these considerations are enough to undermine the notion that the uniquely defined hypersurfaces in the Malament world are connected with simultaneity in the ordinary sense of the term. Yet the arguments of Friedman and Torretti seem to apply even in this case, and this suggests that something important is missing in their analysis, and indeed in the very motivation of Grunbaum's initial problem. (In fact it is even worse than this. To have time appearing in the world, we need more structure, but adding just one more world-line into the Malament world prevents simultaneity-relative-to-W from being uniquely defined.[36]) The missing insight was, to repeat, provided by Poincaré in his analysis of time: we choose a simultaneity relation which optimally simplifies the dynamics. In 1978 C. Giannoni formulated Maxwellian electrodynamics allowing for an arbitrary degree of anisotropy in the one-way speed of light,[37] and one has only to see the complications introduced into Maxwell's equations as a result of this procedure to understand why we standardly choose the isotropic option. The point is that we understand simultaneity not by abstracting away from dynamics but by facing its full complexity.

A final points about the logic of the Friedman–Torretti argument. There is a clear admission, particularly in the passage from Friedman above, that it is

A more systematic discussion of what is aptly called 'modal mayhem' in a somewhat different context (speculations about the physics of worlds with spatial dimensions other than 3) is found in Callender (2005), §5.3.

[36] The realization that the addition of extra structure to the Malament world, even of a rather limited kind, can nullify the Malament result is evident in Janis (1983) and in a different context in Budden (1997*b*).

[37] Giannoni (1978).

not impossible to define a non-standard simultaneity relation: it is just that to characterize it in terms of geometric, coordinate-independent structure something more needs to be added to the two given elements in the Malament world, or indeed to Minkowski space-time with a single world-line.

Let us put aside for the moment the point made above that imposing extra structure is *inevitable* if time in the ordinary sense is to emerge. Another way of describing Friedman's appeal to the principle of parsimony in this context is to say that considerations of simplicity are in play. But does Friedman's reasoning differ fundamentally in form from the usual argument conventionalists use to justify the Poincaré–Einstein convention? The issue is no different from that of choosing spherical polar coordinates when the dynamical system under investigation has spherical symmetry. Friedman's logic, though wrapped up in coordinate-free language, seems in essence very much like the traditional one. It is hard then to see why the traditional approach to the conventionalism issue is 'trivial', but the coordinate-independent one isn't.

6.3.2 The Edwards–Winnie Synchrony-general Transformations

For the sake of simplicity let's restrict space-time to a single space and a time dimension, and assign the Reichenbach synchrony factor ϵ to the frame S, and ϵ' to the frame S'. Then if the Lorentz transformations hold between S and S' for the (Poincaré–Einstein, or standard) choice $\epsilon = \epsilon' = 1/2$, it can be shown that in the general case

$$x' = \frac{c}{\sqrt{A}}(x - v_\epsilon t) \tag{6.5}$$

$$t' = \frac{1}{c\sqrt{A}}(Yt - Xx), \tag{6.6}$$

where c is the two-way light speed and

$$A = c^2 + 2(1 - 2\epsilon)cv_\epsilon - 4\epsilon(1 - \epsilon)v_\epsilon^2 \tag{6.7}$$

$$X = 2(\epsilon - \epsilon')c + 4\epsilon(1 - \epsilon)v_\epsilon \tag{6.8}$$

$$Y = c^2 + 2(1 - \epsilon - \epsilon')cv_\epsilon. \tag{6.9}$$

These transformations were introduced by Edwards in 1967 and further studied by Winnie in 1970.[38] Some features of the Edwards–Winnie transformations are worth emphasizing.

[38] See Edwards (1963) and Winnie (1970).

(i) The condition of Reciprocity (recall that this is the claim that the speed of S' relative to S is the negative of the speed of S relative to S') holds if and only if $\epsilon = \epsilon' = 1/2$.[39] Actually, this result holds even when the only constraint on the coordinate transformations is satisfaction of the *synchrony-independent* version of the invariance of the light-speed, or what Winnie called the Round Trip Light Principle (RTLP). This is the claim that the average round trip speed of any light signal propagating in a closed path is equal to a constant c relative to all inertial frames. The coordinate transformations consistent with RTLP are more general than (6.5–6.6) and take the form

$$x' = \frac{1}{\mathcal{C}_{\|}}(x - v_\epsilon t) \tag{6.10}$$

$$t' = \frac{1}{c^2\mathcal{C}_{\|}}(Yt - Xx), \tag{6.11}$$

where X and Y are defined as above. It is easy to show that Reciprocity $\Longleftrightarrow$ $\epsilon = \epsilon' = 1/2$ follows from (6.10–6.11).[40]

(ii) It follows from RTLP alone that the ratio of the length change factor to the dilation factor takes the form

$$\mathcal{C}_{\|}/\mathcal{D} = A/c^2, \tag{6.12}$$

where A is defined as in (6.7).[41] Suppose we require that both these kinematical factors are isotropic, so that their ratio is an even function of v_ϵ. Then it follows straightforwardly from (6.12) that $\epsilon = 1/2$.[42] This result is hardly surprising, but it demonstrates that the Poincaré–Einstein convention for frame S can be defined not only in relation to the one-way speed of light but also by way of the behaviour of rods and clocks, given RTLP. (Since the length change and dilation factors for S do not depend on ϵ', the synchrony factor in S' is not constrained; see the Sjodin–Tangherlini transformations in point (vi) below.) Note that the notion of isotropy here is quite distinct from the notion of isotropy used in the Poincaré–Einstein derivation of $k = 1$, as outlined in section 5.4.3 above. In our present argument, it is being used to constrain the ϵ factor. In the Poincaré–Einstein derivation, the isotropy of rods and clocks is defined relative to the *pre-chosen* value $\epsilon = 1/2$.

(iii) Winnie showed, without assuming the standard convention $\epsilon = 1/2$, and thus avoiding circularity, that the method of synchronizing clocks by slow clock transport approximates the Poincaré–Einstein convention.[43]

[39] Winnie (1970). [40] See Brown (1990), fn. 5. [41] Brown (1990).

[42] This point was made in Brown and Maia (1993), p. 402.

[43] Winnie (1970). This result had previously been obtained by Eddington in 1923; see Eddington (1965), §§4, 11. See also Torretti (1983), pp. 226–7.

(iv) For $\epsilon \neq \epsilon'$, the Edwards–Winnie transformations (6.5–6.6) do not form a group when $\epsilon \in [0, 1]$. Rather than see this as a grave defect, it should remind us that associating group structure with symmetries (in this case the relativity principle) is perhaps more subtle than meets the eye. But the transformations do form a group when the parameter is invariant.

(v) Suppose we have a homogeneous, istotropic body of air that is at rest relative to S. Choose the Poincaré–Einstein convention in S (relative to which both light and sound propagate isotropically in a one-way sense) and the convention in S' making *only sound* propagate isotropically. If ω is the isotropic one-way speed of sound relative to S, then $(-\omega)' = -(\omega')$. This implies that

$$x' = \gamma(x - vt) \tag{6.13}$$

$$t' = \frac{1}{\gamma(1 - v^2/\omega^2)}(t - vx/\omega^2), \tag{6.14}$$

where γ is the usual Lorentz factor. The contraction and dilation effects are the usual relativistic ones (since these only depend on the choice of ϵ) but the relativity of simultaneity factor takes on, as expected, a new form: $\alpha = v/\omega^2$. This acoustic formulation of relativistic kinematics, due to Michael Redhead,[44] corresponds to the choices

$$\epsilon = 1/2 \tag{6.15}$$

$$\epsilon' = 1/2 + \frac{v(c^2 - \omega^2)}{2c(v^2 - \omega^2)}. \tag{6.16}$$

This contrivance was devised to yield the so-called Zahar transformations in the non-relativistic limit.[45]

(vi) The absence of the relativity of simultaneity ($\alpha = 0$ hence $X = 0$) is equivalent to the condition

$$\epsilon' = \epsilon + 2\epsilon(1 - \epsilon)v_\epsilon/c. \tag{6.17}$$

So

$$x' = \frac{c}{\sqrt{A}}(x - v - \epsilon t) \tag{6.18}$$

$$t' = \frac{\sqrt{A}}{c}t. \tag{6.19}$$

[44] Redhead (1983). [45] Zahar (1977).

Putting $\epsilon = \epsilon'$, we have two possibilities: $\epsilon = \epsilon' = 0$ and $\epsilon = \epsilon' = 1$. According to the first

$$x' = (1 + 2v/c)^{-1/2}(x - vt) \tag{6.20}$$

$$t' = (1 + 2v/c)^{1/2}t. \tag{6.21}$$

According to the second

$$x' = (1 - 2v/c)^{-1/2}(x - vt) \tag{6.22}$$

$$t' = (1 - 2v/c)^{1/2}t. \tag{6.23}$$

Putting $\epsilon = 1/2$ and $\epsilon' = (c + v)/2c$, we obtain the *Sjödin–Tangherlini transformations*:[46]

$$x' = \gamma(x - vt) \tag{6.24}$$

$$t' = t/\gamma. \tag{6.25}$$

Note first that in transformations (6.20)–(6.21) and (6.22)–(6.23) the length change and dilation factors are not even functions of v and hence are anisotropic. This anisotropy is merely an artefact of the choice of the non-standard synchrony factors, and is quite distinct in nature from the objective anisotropy we encountered in the Bogoslovsky–Budden transformations in section 5.4.3. Second, the Sjödin–Tangherlini transformations (6.24)–(6.25) demonstrate that the ϵ-related anisotropy is not a necessary feature of relativistic kinematics constructed to exclude relativity of simultaneity. Admittedly, such exclusion is obtained by means of an artificial procedure, but what all these transformations consistent with (6.18)–(6.19) show is that *explanations of synchrony-independent phenomena in SR that rely crucially on the relativity of simultaneity are not fundamental.* (A common example concerns the clock retardation effect, or 'twins paradox', where it is claimed that at the point of turn-around of the travelling clock, the hyperplanes of simultaneity suddenly change orientation and the resulting 'lost time' accounts for the fact that the clocks when reunited are out of phase. It is worth bearing in mind that the clock retardation effect, like any other synchrony-independent phenomenon in SR, is perfectly consistent with all the non-standard transformations in this section, including those which eliminate relativity of simultaneity.[47])

6.4 RELAXING THE LIGHT POSTULATE: THE IGNATOWSKI TRANSFORMATIONS

An idea that has appeared a number of times in the literature on relativistic kinematics is that of exploring the consequences of dropping Einstein's light postulate

[46] Sjödin (1979). [47] For further discussion of this point, see Debs and Redhead (1996).

in his derivation of the Lorentz transformations. The first case appears to be due to W. von Ignatowski, a Russian mathematical physicist, in work that appeared in 1910 and 1911.[48] Here is a reconstruction of the argument.

Let's recall (from section 3.4) the general linear transformations between frames S and S', whose adapted coordinate systems are in the standard configuration

$$x' = \frac{1}{\mathcal{C}_{\parallel}}(x - vt) \tag{6.26}$$

$$y' = \frac{1}{\mathcal{C}_{\perp}}y \tag{6.27}$$

$$z' = \frac{1}{\mathcal{C}_{\perp}}z \tag{6.28}$$

$$t' = \frac{1}{\mathcal{D}(1 - \alpha v)}(t - \alpha x). \tag{6.29}$$

Recall also that it follows from (6.26) and (6.29) that the velocity transformation rule for a signal moving along the common x, x'-axis is

$$u' = \frac{\mathcal{D}(1 - \alpha v)}{\mathcal{C}_{\parallel}} \frac{(u - v)}{(1 - \alpha u)}. \tag{6.30}$$

Step 1

We would like to resort to something like the Poincaré–Einstein argument (outlined in section 5.4.3) based on joint application of the Relativity Principle (RP) and spatial isotropy to conclude that

$$\mathcal{C}_{\perp} = 1. \tag{6.31}$$

But recall that the original argument is given in the context of the Poincaré–Einstein convention for synchronizing clocks. But now we have no fundamental speed to exploit in this sense, so we just have to assume that there exists a method of synchronizing clocks—one which can be applied in any inertial frame in the spirit of RP—in terms of which the isotropy of the length-change factors manifests itself.

Step 2

Ignatowski erroneously assumed that Reciprocity is a consequence of the RP.[49] A counter-example is provided by the Edwards–Winnie transformations for any

[48] References to the relevant papers of Ignatowski can be found in Torretti (1983); later rediscoveries of his approach to kinematics are found in Lee and Kalotas (1975) and Lévy-Leblond (1976).

[49] Ignatowski argued that Reciprocity is a consequence of the equal passage times principle. This states that rods of equal lengths at rest relative to S and S' take equal times to pass a clock at rest at, say, the origin of the frame moving relative to the rod. This principle would reappear later in

choice of $\epsilon = \epsilon' \neq 1/2$. The transformations form a group, but violate Reciprocity (see points (i) and (iv) in section 6.2). However, in 1969, Berzi and Gorini proved that the combination of RP and spatial isotropy does imply Reciprocity. The proof is somewhat elaborate, and I have not been able to simplify it.[50] I shall assume at any rate that if we apply the synchronization convention introduced in Step 1 to both frames S and S' then Reciprocity will hold. Recall that Reciprocity holds only if

$$\mathcal{D}(1 - \alpha v) = \mathcal{C}_{\|}. \tag{6.32}$$

Equations (6.31) and (6.32) thus lead to what might be called the *Reciprocity Transformations*

$$x' = \frac{1}{\mathcal{C}_{\|}}(x - vt) \tag{6.33}$$

$$y' = y \tag{6.34}$$

$$z' = z \tag{6.35}$$

$$t' = \frac{1}{\mathcal{C}_{\|}}(t - \alpha x), \tag{6.36}$$

with the associated (longitudinal) velocity transformation

$$u' = \frac{(u - v)}{(1 - \alpha u)}. \tag{6.37}$$

Step 3

We have already applied the RP to the transverse transformations (6.27), (6.28) in Step 1. We now do likewise to the transformations (6.33) and (6.36). First we simply algebraically invert these equations:

$$x = \frac{\mathcal{C}_{\|}(v)}{(1 - \alpha(v)v)}(x' + vt') \tag{6.38}$$

$$t = \frac{\mathcal{C}_{\|}(v)}{(1 - \alpha(v)v)}(t' + \alpha(v)x'). \tag{6.39}$$

Winnie (1970). Both Ignatowski and Winnie incorrectly thought that the principle is a consequence of the RP. In Ignatowski's case, the error was pointed out by Torretti (1983), fn. 5, pp. 79, 298; in Winnie's case by Brown (1990). A careful treatment of the Ignatowski derivation as a whole is found in Torretti (1983), pp. 76–82.

[50] See Berzi and Gorini (1969). For a summary of the proof, see Torretti (1983), pp. 79–80, fn. 7, p. 298, but see a correction in Budden (1997*b*). The attempt by Rindler to establish reciprocity just on the basis of isotropy is criticized in Brown and Maia (1993), Appendix II.

Now the RP implies that these equations should have the same form as (6.33) and (6.36), i.e.

$$x = \frac{1}{\mathcal{C}_{\|}(-v)}(x' + vt') \tag{6.40}$$

$$t = \frac{1}{\mathcal{C}_{\|}(-v)}(t' - \alpha(-v)x'). \tag{6.41}$$

So equating (6.38) with (6.40), and (6.39) with (6.41), we obtain (recalling from Step 1 that in the chosen synchrony convention $\mathcal{C}_{\|}(-v) = \mathcal{C}_{\|}(v)$)

$$\mathcal{C}_{\|}(v) = (1 - \alpha(v)v)^{1/2} \tag{6.42}$$

$$\alpha(-v) = -\alpha(v), \tag{6.43}$$

yielding

$$x' = (1 - \alpha v)^{-1/2}(x - vt) \tag{6.44}$$

$$y' = y \tag{6.45}$$

$$z' = z \tag{6.46}$$

$$t' = (1 - \alpha v)^{-1/2}(t - \alpha x). \tag{6.47}$$

Note that for the coordinates x' and t' to remain finite, we must have $|1-\alpha v| > 0$.

Step 4

Now consider a further transformation from frame S' to S'', where S' is also in the standard configuration and moving with velocity v' relative to S'. Using $u' = v'$ and solving for u in (6.37), it is easily seen that the velocity w of S'' relative to S is given by

$$w = \frac{v + v'}{1 + \alpha(v)v'}. \tag{6.48}$$

For simplicity let us write $(1 - \alpha v)^{-1/2} = \mu(v)$. Then using RP and hence the form of (6.44) to (6.47) for *each* of the transformations $S \to S'$ and $S' \to S''$, and combining, we obtain the transformations for $S \to S''$:

$$x'' = \mu(v')\mu(v)(1 + \alpha(v)v')^{-1/2}(x - wt) \tag{6.49}$$

$$y'' = y \tag{6.50}$$

$$z'' = z \tag{6.51}$$

$$t'' = \mu(v')\mu(v)(1 + \alpha(v')v)^{-1/2}(t - \hat{\alpha}x), \tag{6.52}$$

where

$$\hat{\alpha} = \frac{\alpha(v) + \alpha(v')}{1 + \alpha(v')v}. \tag{6.53}$$

The transformations (6.49–6.52) should in turn take the same form as (6.44–6.47). This means that for all v, v'

$$\alpha(v)v' = \alpha(v')v \tag{6.54}$$

or

$$\frac{\alpha(v)}{v} = \frac{\alpha(v')}{v'} = K, \tag{6.55}$$

where K is a universal constant of dimensions vel^{-2}. (This is consistent with $\hat{\alpha} = \alpha(w)$.)

So we obtain finally the *Ignatowski transformations*

$$x' = (1 - Kv^2)^{-1/2}(x - vt) \tag{6.56}$$

$$y' = y \tag{6.57}$$

$$z' = z \tag{6.58}$$

$$t' = (1 - Kv^2)^{-1/2}(t - Kvx), \tag{6.59}$$

where $|v| < K^{-1/2}$.

6.4.1 Comments

(i) No assumption about the existence of an invariant speed entered into the proof of the Ignatowski transformations. However, it is easily seen that the transformations themselves imply the existence of an invariant speed: $K^{-1/2}$.

(ii) Putting $K = 0$ (invariant speed is infinite) we obtain the Galilean transformations.

(iii) Putting $K > 0$ and simply writing $c = K^{-1/2}$, we obtain the Lorentz transformations. But we have not obtained relativistic kinematics in the usual sense of the term. For the physical connection between $K^{-1/2}$ and the light speed has not been established.

(iv) The derivation of the Ignatowski transformations yields the isotropy of the one-way invariant speed $K^{-1/2}$, on the basis of the synchrony convention associated with steps 1 and 2. This seems to mean that the assumption of (one-way) isotropy of space deformation and time dilation, together with *either* the RP (given the results of Ignatowski, and Berzi and Gorini) *or* the Round Trip Light Principle (invariance of the two-way light speed) as seen in point (ii) in section 6.3.2, is sufficient to guarantee that the invariant speed is isotropic. (This point should not be interpreted as a establishing that the issue of spreading time through space is anything other than based on convention!)

6.5 THE NON-RELATIVISTIC LIMIT

Formally, to go from the Lorentz transformations to the Galilean transformations we can take the limit $c \rightarrow \infty$. It corresponds to going from a finite value of K to the zero value in the Ignatowski transformations. But this is not the right way to think about the non-relativistic limit of relativistic kinematics.

The reason is simple: in our world c, the invariant speed, is finite. What we want is to specify conditions that pertain to our world and under which non-relativistic kinematics emerge to a high degree of approximation. In a penetrating 1989 treatment of the reduction of general relativity to the Newtonian theory of gravity, F. Rohrlich compared what he called the 'dimensionless' process of theory reduction to the 'dimensional' one.[51] The former generally takes suitable dimensionless quantities—the ratio of two physical quantities of the same dimensions—to be negligibly small. The latter involves taking limits of dimensional parameters such as the light speed c or Planck's constant $\hbar$. Rohrlich emphasized that the dimensionless process represents a case of 'factual' approximation and that the dimensional approximation is 'counterfactual', because for instance it is a fact that c is finite. What we are interested in here is the factual approach.

So imagine two frames S and S' whose adapted coordinates are in the standard configuration and transform according to the Lorentz transformations (5.4–5.7). Let us suppose first that we consider only relative speeds v that satisfy the condition $|v/c| \ll 1$, so that $\gamma \sim 1$. We thus obtain

$$\mathbf{x}' \sim \mathbf{x} - \mathbf{v}t; \quad t' \sim t - \mathbf{v} \cdot \mathbf{x}/c^2. \tag{6.60}$$

A similar result is obtained for the limit of the differential operators

$$\nabla' \sim \nabla + \frac{\mathbf{v}}{c^2}\frac{\partial}{\partial t}, \quad \frac{\partial}{\partial t'} \sim \frac{\partial}{\partial t} + \mathbf{v} \cdot \nabla. \tag{6.61}$$

Note that the temporal transformation in (6.60) incorporates discernible relativity of simultaneity as long as distances are big enough. Indeed, if we have $|\mathbf{x}| \gg ct$, we obtain

$$\mathbf{x}' \sim \mathbf{x}; \quad t' \sim t - \mathbf{v} \cdot \mathbf{x}/c^2, \tag{6.62}$$

[51] Rohrlich (1989).

which represents not a boost but a shift in the simultaneity hyperplanes relative to the original frame S. If we are hoping to get Galilean kinematics, we need to impose the opposite constraint: $|\mathbf{x}| \ll ct$, yielding the desired transformations

$$\mathbf{x}' \sim \mathbf{x} - \mathbf{v}t; \quad t' \sim t. \tag{6.63}$$

However, the Galilean transformations of the differential operators

$$\nabla' \sim \nabla, \quad \frac{\partial}{\partial t'} \sim \frac{\partial}{\partial t} + \mathbf{v} \cdot \nabla \tag{6.64}$$

follow from (6.61) only if the derivatives of relevant functions are such that expressions involving the time-derivative term appearing in the first equation in (6.61) are negligible in this approximation. The feasibility of implementing this condition, which we shall denote as $\nabla \gg (\mathbf{v}/c^2)\partial/\partial t$, depends on the context and requires considerable care.

The conditions imposed on the relativistic kinematics so far do not suffice to regain the full Galilean kinematics because the relativistic velocity transformation rule still differs from the Galilean one if the velocity of the body in question—not to be conflated with the relative velocity $\mathbf{v}$ of the frames—is relativistic. (The Fresnel drag coefficient which follows from this rule is measurable even when $|\mathbf{v}|/c \ll 1$ because the object is light itself.) Thus we need further to assume that all bodies move at non-relativistic speeds, so that

$$\dot{\mathbf{x}}' \sim \mathbf{x} - \mathbf{v}. \tag{6.65}$$

Thus we may collect the conditions necessary for Galilean kinematics to emerge:[52]

$$|\mathbf{v}/c| \ll 1; \quad |\mathbf{x}| \ll ct; \quad \nabla \gg (\mathbf{v}/c^2)\partial/\partial t; \quad |\dot{\mathbf{x}}/c| \ll 1. \tag{6.66}$$

But there is a last twist in the story. The Galilean transformations (6.63) are formally right (or would be if they were exact) but what guarantees that the way time is spread through space corresponds to the synchrony convention in Newtonian mechanics? It is here that the Eddington–Winnie theorem, mentioned in section 6.3.2, shows its worth. For it establishes that the Poincaré–Einstein synchrony convention in S is equivalent to that of slow clock transport, and that is all we need.

A final word concerns Maxwellian electrodynamics, the first relativistic theory. It has been known for some time that this theory has two non-relativistic limits, corresponding to the striking fact that deleting one of either two terms in

[52] For a fuller discussion see Holland and Brown (2003).

Maxwell's equations—one corresponding to the Faraday induction term, and the other corresponding to the displacement current—renders the equations Galilean covariant. (Note that this fact does not in itself imply that conditions can be found for real electric and magnetic fields that yield the Galilean-covariant behaviour in the appropriate approximation.) That even one limit exists is noteworthy given that the theory is one of a massless field, although in this sense the situation is no different from the case of gravity. When Maxwell's equations are coupled with the Dirac equation, however, there appears to be a single non-relativistic limit.[53]

[53] The first systematic treatment of the non-relativistic limit of Maxwell's equations to our knowledge was given by Bellac and Lévy-Leblond (1973); a somewhat different treatment is found in Holland and Brown (2003), from which the conditions (6.66) are taken, and which also discusses the non-relativistic limit of the coupled Maxwell–Dirac equations.

7

Unconventional Voices on Special Relativity

I think it may be pedagogically useful to start with the example, integrating the equations in some pedestrian way, for example by numerical computation. The general argument, involving as it does a change of variables, can (I fear) set off premature philosophizing about space and time.

John S. Bell[1]

7.1 EINSTEIN HIMSELF

A theme running through Einstein's writings was that what he called 'elementary' foundations were unavailable to account for the stable structure and cohesion of matter, and that this was the reason he constructed SR in the way he did. We saw in section 5.2 that as early as 1908, he was referring to SR as a 'half' salvation, given its inspiration in thermodynamics. In the 1919 article for the London *Times* where he made the principle-constructive theory distinction, he stated quite clearly that 'when we say we have succeeded in understanding a group of natural processes, we invariably mean that a constructive theory has been found which covers the processes in question.'

But the SR of 1905 is not a constructive theory, and part of the reason is the role that rods and clocks play in Einstein's attempt to operationalize inertial coordinate systems. It is evident that Einstein harboured a sense of unease about their status in his theory. In his 1921 essay entitled 'Geometry and Experience', he wrote:

It is . . . clear that the solid body and the clock do not in the conceptual edifice of physics play the part of irreducible elements, but that of composite structures, which must not play any independent part in theoretical physics. But it is my conviction that in the present stage of development of theoretical physics these concepts must still be employed as independent concepts; for we are still far from possessing such certain knowledge of the theoretical principles of atomic structure as to be able to construct solid bodies and clocks theoretically from elementary concepts.[2]

[1] Bell (1976*a*), fn. 10. [2] Einstein (1921).

Einstein's unease is again clearly expressed in a similar passage in his 1949 *Autobiographical Notes*:

One is struck [by the fact] that the theory [of special relativity] . . . introduces two kinds of physical things, i.e. (1) measuring rods and clocks, (2) all other things, e.g., the electromagnetic field, the material point, etc. This, in a certain sense, is inconsistent; strictly speaking measuring rods and clocks would have to be represented as solutions of the basic equations (objects consisting of moving atomic configurations), not, as it were, as theoretically self-sufficient entities. However, the procedure justifies itself because it was clear from the very beginning that the postulates of the theory are not strong enough to deduce from them sufficiently complete equations . . . in order to base upon such a foundation a theory of measuring rods and clocks. . . . But one must not legalize the mentioned sin so far as to imagine that intervals are physical entities of a special type, intrinsically different from other variables ('reducing physics to geometry', etc.).[3]

These remarks are noteworthy for several reasons.

First, there is the issue of justifying the 'sin' of treating rods and clocks as primitive, or unstructured entities in SR. Einstein does not say, as he does in 1908 and 1921, that the 'elementary' foundations of a constructive theory of matter are still unavailable; rather he simply reminds us of the limits built into the very form of the 1905 theory. It is hardly any justification at all. Progress in the relativistic quantum theory of matter had been made between 1905 and 1949. Was it Einstein's long-standing distrust of the quantum theory that held him back from recognizing this progress and its implications for his formulation of SR?[4]

Second, it is clear that not even Einstein ever 'fully made the transition from the old dynamics to the new kinematics' (recall, in the Preface above, Abraham Pais's criticism of Lorentz for having failed to make this transition). For to say that length contraction is intrinsically kinematical would be like saying that energy or entropy are intrinsically thermodynamical, not mechanical—something Einstein would never have accepted.

Third, there is a hint at the end of the passage that towards the end of his life Einstein still did not view geometrical notions as fundamental in the special theory. We will return to this delicate issue shortly. In the meantime, I shall attempt to outline the doubts raised by some other voices in the twentieth century about the standard treatment of rods and clocks in SR.[5]

7.2 1918: HERMANN WEYL

In discussing the significance of the MM experiment in his 1918 text *Raum-Zeit-Materie*, Hermann Weyl stressed that the null result is a consequence of the fact

[3] Einstein (1969), pp. 59, 61.

[4] Further discussion of Einstein's 1949 acknowledgement of his 'sinful' treatment of rods and clocks is found in Brown and Pooley (2001).

[5] Sections 8.2 and 8.5 of this chapter rely heavily on Brown and Pooley (2001).

that 'the interactions of the cohesive forces of matter as well as the transmission of light' are consistent with the requirement of Lorentz covariance. Weyl's emphasis on the role of 'the mechanics of rigid bodies' in this context indicates a clear understanding of the dynamical underpinnings of relativistic kinematics.[6] But Weyl's awareness that rigid rods and clocks are structured dynamical entities led him to the view that it is wrong to define the 'metric field' in SR on the basis of their behaviour.

Weyl's concern had to do with the problem of accelerated motion, or with deviations from what he called 'quasi-stationary' motion. Weyl's opinion in *Raum-Zeit-Materie* seems to have been that if a clock, say, is undergoing non-inertial motion, then it is unclear in SR whether the proper time read off by the clock is directly related to the length of its world-line determined by the Minkowski metric. For Weyl, clarification of this issue can only emerge 'when we have built up a **dynamics** based on physical and mechanical laws'.[7] In a sense Weyl was right. The claim that the length of a specified segment of an arbitrary time-like curve in Minkowski spacetime, obtained by integrating the Minkowski line element ds along the segment, is just the clock hypothesis we met in section 6.2.1. Recall that its justification rests on the contingent dynamical requirement that the external forces accelerating the clock are small in relation to the internal 'restoring' forces at work inside the clock. This dynamical theme was to re-emerge in Weyl's responses to Einstein's criticisms of his 1918 attempt at a unified field theory.

Weyl's 1918 publication[8] of a stunning, though doomed, unification of gravitational and electromagnetic forces raised a number of intriguing questions about the meaning of space-time structure. Weyl started from the claim that the pseudo-Riemannian space-time geometry of Einstein's general relativity is not sufficiently local in that it allows the comparison of the lengths of distant vectors. Instead, Weyl insisted that the choice of unit of (space-time) length at *each* point is arbitrary: only the ratios of the lengths of vectors *at the same point* and the angles between them can be physically meaningful. Such information is invariant under a *gauge transformation* of the metric field: $g_{ij} \longrightarrow g'_{ij} = e^{2\lambda(x)} g_{ij}$ and constitutes a conformal geometry.

In addition to this conformal structure, Weyl postulated that space time is equipped with an affine connection that preserves the conformal structure under infinitesimal parallel transport. In other words, the infinitesimal parallel transport of all vectors at p to p' is to produce a similar image at p' of the vector space at p.[9] For a given choice of gauge, the constant of proportionality of this similarity mapping will be fixed. Weyl assumed that it differed infinitesimally from unity and thereby proceeded to show that the coefficients of the affine connection depended

[6] Weyl (1952), pp. 173. [7] op. cit., p. 177. [8] Weyl (1918).

[9] It is worth noting at this point that Weyl could, and perhaps should have gone further! As the keen-eyed Einstein was to point out, it is in the spirit of Weyl's original geometric intuition to allow for the relation between tangent spaces to be a weaker affine mapping: why insist that it be a similarity mapping? Einstein made this point in a letter to Weyl in 1918. For details see Vizgin (1985).

on a one-form field ϕ_i in addition to the metric coefficients g_{ij} in such a way that the change in any length l under parallel transport from p (coordinates $\{x^i\}$) to p' (coordinates $\{x^i + dx^i\}$) is given by:

$$dl = l\phi_i dx^i. \tag{7.1}$$

Under the gauge transformation, $g_{ij} \rightarrow g'_{ij} = e^{2\lambda}g_{ij}$, $l \rightarrow e^{\lambda}l$. Substituting this into (7.1) gives:

$$\phi_i \rightarrow \phi'_i = \phi_i + \frac{\partial\lambda}{\partial x^i}, \tag{7.2}$$

the familiar transformation law for the electromagnetic four-potential. Weyl thus identified the gauge-invariant, four-dimensional curl of the geometric quantity ϕ_i with the familiar electromagnetic field tensor.

For a given choice of gauge a comparison of the length of vectors at distant points can be effected by integrating (7.1) along a path connecting the points. This procedure will in general be path independent just if the electromagnetic field tensor vanishes everywhere.

As is well known, despite his admiration for Weyl's theory, Einstein was soon to spot a serious difficulty with the non-integrability of length[10]. In the case of a static gravitational field, a clock undergoing a round-trip in space during which it encountered a spatially varying electromagnetic potential would return to its starting point ticking at a rate different from that of a second clock which had remained at the starting point and which was originally ticking at the same rate. An effect analogous to this 'second clock effect' would occur for the length of an infinitesimal rod under the same circumstances. But it is a fact of the world—and a highly fortunate one!—that the relative periods of clocks (and the relative lengths of rods) do not depend on their relative histories in this sense.

Before looking at Weyl's reply to this conundrum, it is worth remarking that it was apparently only in 1983 that the question was asked: what became of Einstein's objection once the gauge principle found its natural home in quantum mechanics? C. N. Yang pointed out that because the non-integrable scale factor in quantum mechanics relates to phase, the second clock effect could be detected using wave-functions rather than clocks, essentially what Yakir Aharonov and David Bohm had discovered.[11]. Note that Yang's question can be inverted: is there a full analogue of the Aharonov–Bohm effect in Weyl's gauge theory?[12] The answer is yes, and it indicates that there was a further sting in Einstein's objection to Weyl that he and his contemporaries failed to spot. The point is that *the second-clock effect*

[10] Einstein (1918*a*).
[11] See Yang (1984), Aharonov and Bohm (1959). Yang recounts this incident in Yang (1986).
[12] This point, as with much of this section, is taken from Brown and Pooley (2001).

obtains in Weyl's theory even when the electromagnetic field vanishes everywhere on the trajectory of the clock, so long as the closed path of the clock encloses some region in which there is a non-vanishing field. This circumstance highlights the difficulty one would face in providing a dynamical or 'constructive' account of the second-clock effect in Weyl's theory.[13] The theory seems to be bedevilled by non-locality of a very striking kind.

The precise nature of Weyl's responses to Einstein's objection would vary in the years following 1918 as he went on to develop new formulations of his unified field theory based on the gauge principle. But the common element was Weyl's rejection of the view that the metric field could be assigned operational significance in terms of the behaviour of rods and clocks. His initial argument was an extension of the point he made about the behaviour of clocks in SR: one cannot know how a clock will behave under accelerations and in the presence of electromagnetic fields until a full dynamical modelling of the clock under these circumstances is available. The price Weyl ultimately paid for the beauty of his gauge principle—quite apart from the complicated nature of his field equations—was the introduction of tentative and ultimately groundless speculations concerning a complicated dynamical adjustment of rods and clocks to the 'world curvature' so as to avoid the second-clock effect and its analogue for rods.

We finish this section with a final observation on the nature of Weyl's theory, with an eye to issues in standard general relativity to be discussed later. It was noted above that Weyl's connection is not a metric connection. It is a function not only of the metric and its first derivatives, but also depends on the electromagnetic gauge field: in particular, for a fixed choice of gauge, the covariant derivative of the metric does not vanish everywhere. What does this imply?

The vanishing of the covariant derivative of the metric—the condition of metric compatibility—is sometimes introduced perfunctorily in texts on general relativity, but Schrödinger was right to call it 'momentous'.[14] It means that the local Lorentz frames associated with a space-time point p (those for which, at p, the metric tensor takes the form $\mathrm{diag}(1, -1, -1, -1)$ and the first derivatives of all its components vanish) are also local inertial frames (relative to which the components of the connection vanish at p).[15] If the laws of physics of the non-gravitational interactions are assumed to take their standard special relativistic form at p relative to such local Lorentz charts (the local validity of special relativity), then metric compatibility implies that gravity is not a force in the

[13] It is worth noting that in 1923, Lorentz himself wrote, in relation to the rod analogue of the second clock effect in the Weyl theory, that this 'would amount to an action of an electromagnetic field widely different from anything that could reasonably be expected'; see Lorentz (1937). But whether Lorentz was concerned with the dynamical problem of accounting for how Maxwell's electrodynamics could in principle have such an effect on physical bodies like rods—a consideration which one would not expect to be foreign to Lorentz's thinking!—or simply with the empirical fact that such an effect is non-existent, is not entirely clear from Lorentz's comments.

[14] Schrödinger (1985), p. 106.

[15] See Misner *et al.* (1973), p. 313.

traditional sense—an agency causing deviation from natural motion—insofar as the worldlines of freely falling bodies are geodesics of the connection.

The full physical implications of the non-metric compatible connection in Weyl's theory remain obscure in the absence of a full-blown theory of matter. Weyl's hints at a solution to the Einstein objection seem to involve a violation of minimal coupling, i.e. a violation of the prohibition of curvature coupling in the non-gravitational equations, and hence of the local validity of special relativity. It seems that the familiar insight into the special nature of the gravitational interaction provided by the strong equivalence principle—the encapsulation of the considerations given in the previous paragraph—is lost in the Weyl theory.

7.3 1920S: PAULI AND EDDINGTON

The acclaimed 1921 review article on relativity theory, special and general, by Wolfgang Pauli (1900–1958) is one of the supreme examples of precociousness in the history of physics: Pauli was 21 when it was published, in the same year that he completed his doctoral thesis on the quantum theory of ionized molecular hydrogen. The maturity of this review, and the sure-footedness of its historical parts, are astonishing.[16] In the section of the essay discussing Einstein's 1905 paper, Pauli was struck by the 'great value' of the apparent fact that, unlike Lorentz, Einstein had given an account of his kinematics which was free of assumptions about the constitution of matter. He wrote:

> Should one, then, completely abandon any attempt to explain the Lorentz contraction atomistically? We think that the answer to this question should be No. The contraction of a measuring rod is not an elementary but a very complicated process. It would not take place except for the covariance with respect to the Lorentz group of the basic equations of electron theory, as well as of those laws, as yet unknown to us, which determine the cohesion of the electron itself.

I cannot help wondering if this view was not influenced by the fact that Pauli's first research papers concerned the 1918 Weyl theory.

In his 1928 book *The Nature of the Physical World*, Arthur S. Eddington (1882–1944) attributed length contraction to the electromagnetic nature of the

[16] See Pauli (1981). Here is Einstein's evaluation of the essay:

> Whoever studies this mature and grandly conceived work might not believe that its author is a twenty-one year old man. One wonders what to admire most, the psychological understanding for the development of ideas, the sureness of mathematical deduction, the profound physical insight, the capacity for lucid, systematical presentation, the knowledge of the literature, the complete treatment of the subject matter, or the sureness of critical appraisal. (Quoted in the essay 'Wolfgang Ernst Pauli', in Pais (2000), pp. 210–62.)

composition of a rigid rod:

There is really nothing mysterious about the FitzGerald contraction. It would be an unnatural property of a rod pictured in the old way as continuous substance occupying space in virtue of its substantiality; but it is an entirely natural property of a swarm of particles held in delicate balance by electromagnetic forces, and occupying space by buffeting away anything that tries to enter.[17]

Eddington's informal clarification of the nature of the contraction is a little off-beam. He puts emphasis on the appearance of magnetic fields inside the moving rod, whose particles at rest assume a rigid configuration as a result only of electric forces. Eddington both attributes a stress to this magnetic effect, and does not seem to be aware of the distortion produced by motion in the electric forces. Moreover, he makes no mention of the effects of motion on the non-electromagnetic forces involved in the cohesion of matter. At any rate, for a quantitative analysis he misleadingly points the reader in the direction of the writings of Lorentz and Larmor of around 1900, in which as we have seen, length contraction is not derived in any systematic fashion. (To be fair to Eddington, he knew something in 1928 that Lorentz and Larmor didn't: that the principal force of cohesion in matter is after all electromagnetic. But this is not the whole story of course, as we see below.) However, Eddington correctly puts the qualitative cause of the contraction squarely in the nature of the forces of cohesion:

It is necessary to rid our minds of the idea that this failure to keep a constant length is an imperfection of the rod; it is only imperfect as compared with an imaginary 'something' which has not this electrical constitution—and therefore is not material at all. The FitzGerald contraction is not an imperfection but a fixed and characteristic property of matter, like inertia.[18]

7.4 1930S AND 1940S: W. F. G. SWANN

As early as 1912, the British physicist William F. G. Swann (1884–1962), best known later for his work in the USA on cosmic rays, stressed that the Lorentz covariance of Maxwell's equations was far from sufficient in accounting for length contraction. In itself this point was far from new; we have seen in Chapter 6 how clearly it was appreciated by Lorentz, Larmor, and Poincaré. But in papers in 1930 and particularly 1941, Swann stressed that the extra ingredient had to be relativistic quantum theory, and the prospect it held out for an explanation

[17] Eddington (1928), p. 7.

[18] op. cit., p. 7. It is noteworthy that Eddington's thinking on the issue appears to have gone through a change. In his earlier books on relativity dating from 1920 and 1924 (specifically Eddington (1966), p. 151 and Eddington (1965), pp. 26–7), the discussions of length contraction present it as a perspectival effect which, at least in the 1920 discussion, is contrasted with the view based on the 'the behaviour of electrical forces'.

of the cohesion of matter. As far as I know, Swann was the first to stress the important role played by quantum theory in understanding the physics of the Lorentz transformations.[19]

Swann was of course aware of the nature of Einstein's derivation of the Lorentz transformations, but seemed to regard it as providing virtually no *explanation* as to why rods in fact contract and clocks dilate.[20] Swann seemed to think (like Einstein in 1949) that the fundamental tenet of SR is the Lorentz-covariance of the laws of physics, and his particular emphasis was on electrodynamics. In his 1941 papers he claimed that there would be meaning to SR even if the FitzGerald–Lorentz contraction were not to hold, and the result of the Michelson–Morley experiment were non-null.

What Swann meant by this strange-sounding claim was this.[21] It is a mathematical fact, having nothing to do with the usual physical meaning of the variables appearing in the Lorentz transformations (and of course the concomitant transformations of the field components) that Maxwell's equations are Lorentz covariant. In itself this fact does not imply that the primed (transformed) variables refer to physical quantities actually measured by an inertial observer moving at the appropriate speed v relative to the unprimed frame. (In fact, Lorentz denied it, as we have seen.) Thus, Swann concludes that SR, which secures the mathematical fact of Lorentz covariance, does not in itself determine how rods and clocks behave in motion. One can imagine Pauli protesting at this point, and reminding us that if *all* the laws of physics are Lorentz covariant, which must also follow from SR, including the laws governing all the forces of cohesion in matter, then length contraction and time dilation are guaranteed. Quite so. There is no doubt Swann overstated his case, but the merit of his argument is that he went on to stress that these laws must be of a quantum nature, and he spells out the logic in unprecedented detail.

Swann imagines a rigid rod at rest relative to the frame S, whose form and stability are determined by the rod being in a quantum mechanical ground state ψ, robust under small perturbations. If the quantum mechanical equations (which must in principle govern all the forces involved in atomic structure, not just electromagnetic) are Lorentz covariant, then we can guarantee the existence of another solution $\tilde{\psi}$, another ground state, but now describing the rod moving

[19] It is known that Swann corresponded with Einstein on the foundations of SR; in Stachel (1986), p. 378, part of a 1942 letter from Einstein to Swann is cited in which he discusses the possibility of a constructive formulation of the theory wherein rods and clocks are not introduced as 'independent objects'. Einstein argues in this letter that any such theory must, like the quantum theory, contain an absolute scale of length. It would be interesting to know more about this correspondence.

[20] Swann (1930), p. 248. This lengthy paper on 'Relativity and Electrodynamics'—over 60 pages—which appeared in the second volume of *Reviews of Modern Physics*, is curious both in content and style; it is long-winded, thoughtful, and somewhat idiosyncratic; and it contains not a single reference other than to the author's own publications! The Editors didn't seem to mind: these features are shared by Swann's two (shorter) 1941 papers in the same journal: Swann (1941*a*) and Swann (1941*b*).

[21] Swann (1941*b*).

with uniform speed v relative to S, and longitudinally Lorentz contracted by the factor γ^{-1}. Remember: a symmetry transformation takes solutions of the equations of motion into new solutions, and for every v we have a new 'moving' solution. It is not necessary to know the precise form of the quantum equations, just that they are Lorentz covariant. So far nothing has been said about co-moving coordinates, but if a co-moving observer adopts coordinates that match the moving rods and clocks associated with the new solution (the same argument can be used for clocks)—recall the crucial assumption 2 discussed in section 2.4 concerning the boostability of rulers and clocks—then the quantum description of the moving rod by the comoving observer is exactly the same as the description of the originally stationary rod by the observer associated with the frame S. The two coordinate systems will be related by a Lorentz transformation once the Einstein convention for synchronizing distant clocks in the moving frame is adopted.

Two points are worth stressing about this argument. First, the relativity principle is a consequence of the Lorentz covariance of the quantum dynamics, rather than the other way round. Second, the assumption 1 in section 2.4 concerning the universality of the behaviour of rods and clocks, which, it will be recalled, is crucial for the coordinate transformations to carry empirical content in terms of length contraction and time dilation, now also emerges as a consequence of the dynamical argument, as long as matter of any constitution is assumed in principle to come under the sway of the quantum theory in question.

Swann's own conclusion was this:

> It thus appears that a relativistically invariant quantum theory, or something closely analogous to it, is a necessary supplement to the general principle of invariance of equations if we are to provide for the Fitzgerald-Lorentz contraction and for the customarily accepted form of the theory of relativity.[22]

But if we ask what the equations are which appear in this general principle of invariance, at the most fundamental level they will already be equations in quantum theory. After all, even the electromagnetic field—the object of Swann's attention—is ultimately quantized. It may be then that Swann's articulation of the main point leaves a little to be desired, but the main point itself stands. Rods and clocks in motion behave relativistically because the forces that are responsible for the cohesion of matter, forces which inevitably are described by quantum theory, satisfy equations that are Lorentz covariant. It is noteworthy that the essence of Swann's message was recently rediscovered by S. Liberati, S. Sonego, and M. Visser. In 2002, they gave with apparent approval the following 'somewhat unusual take on special relatively':

> The Maxwell equations, considered simply as a mathematical system, possess a symmetry, the Lorentz group, under redefinitions of the labels x and t. But this is a purely mathematical statement devoid of interesting consequences until one asks how physical clocks

[22] op. cit., p. 201.

and rulers are constructed and what forces hold them together. Since it is electromagnetic forces balanced against quantum physics which holds the internal structure of these objects together, the experimental observation that to very high accuracy physical bodies also exhibit Lorentz symmetry allows one to deduce that quantum physics obeys the same symmetry as the Maxwell equations. Viewed in this way, all experimental tests of special relativity are really precision experimental tests of the symmetry group of quantum physics.

7.5 1970S: L. JÁNOSSY AND J. S. BELL

7.5.1 L. Jánossy

In 1971 a remarkable book entitled *Theory of Relativity based on Physical Reality* by Lajos Jánossy (1912–1978)[23] was published in Budapest. It contains a careful, technical treatment of relativity theory and its experimental basis from the point of view that relativistic physics, though correct, has 'nothing to do with the structure of space and time'.[24] Care must be exercised in interpreting this claim; the book is essentially an attempt to consolidate and complete Lorentz's programme of providing a dynamical underpinning for relativistic kinematic effects. Unfortunately, Jánossy advocated a notion of the ether as the seat of electromagnetic waves. It turns out to be a very weak notion, and categorically not the basis of a concept of absolute rest. Indeed, it was claimed by Jánossy to be hardly different from the notion of ether that Einstein himself defended in 1924.[25] But one cannot help wondering whether its appearance in the book, which seems unnecessary, did not harm the book's reputation.

Jánossy is clear about the origins of, say, length contraction.

> A solid consists of atoms and the shape of the solid arises as a dynamical equilibrium of these atoms. It must be supposed that the atoms act upon each other in a *retarded* fashion. It can easily be seen that the retarded interaction leads to different equilibrium configurations in the case of atoms at rest and in the case of atoms moving with a constant velocity v.[26]

But the devil is in the details, and Jánossy criticizes Lorentz for appealing to 'independent, more or less accidental circumstances' in his defence of something like this position. Like Einstein, Jánossy wants to provide an account based on a small number of general principles, and the way he goes about it is curious.

Jánossy introduces first the *Lorentz principle.* Imagine a physical system consisting of a collection of moving point particles. A *deformation* of the system

[23] Jánossy, whose foster-father was Georg Lukacs, the Marxist philosopher, emigrated from Hungary in 1919, and received his education in Vienna and Berlin. In 1936 he moved to Manchester to work with P. M. S. Blackett, and in 1947 took up a position in the Institute for Advanced Studies in Dublin. Jánossy returned to his homeland in 1950 at the invitation of the government, and up to 1970 held directorships of a number of laboratories and research institutes in Hungary. (I am grateful to Miklos Redei for these biographical details.)

[24] Jánossy (1971), p. 13.

[25] Einstein (1924).

[26] op. cit., p. 64.

is obtained by transforming the coordinates of the particles according to some subgroup of the Lorentz group, say the proper Lorentz group of boosts. This is not to be interpreted as a passive coordinate transformation, or change of reference system, but rather as an active change in the coordinates of the particles relative to the original inertial coordinate system. The Lorentz principle states that *the laws of physics must be such that the Lorentz deformed system is a possible system obeying the same laws as the original system.*[27] Jánossy goes on to show that the null results of the ether-drift experiments are a consequence of this principle, and that more generally 'from the observations of physical systems we cannot arrive at conclusions as to the translational velocity of these systems relative to the ether provided the internal motion of the system obeys the Lorentz principle.'

The argument is essentially the same as that given by Swann in 1941, except for two important features. First, Jánossy does not bring quantum theory into the picture. In fact, it is initially unclear how the Lorentz principle is supposed to work even in the classical case of a system of interacting particles, or of test particles immersed in a background field, or indeed of just fields. The deformation is defined, as we have seen, in terms of transformations of the coordinates of the particles, but the Lorentz principle would not be expected to hold in general if corresponding transformations in the fields are not also introduced. That this extra element is indeed what Jánossy has in mind only becomes clear later in the book in the discussion of Maxwell's equations and the Lorentz principle.[28] This ambiguity in the formulation of the principle would be removed if Jánossy just equated it with the Lorentz covariance of the fundamental laws of physics, and it is hard to see why he didn't.[29] It is almost as if Jánossy intends the Lorentz principle to stand over Lorentz covariance. At the start of the mentioned discussion of Maxwell's equations, he announces that 'Physically new statements are obtained if we apply the Lorentz principle to Maxwell's equations.' But of course what emerges in the discussion is simply a consequence of the Lorentz covariance of these equations.

The second way in which Jánossy's treatment differs from Swann's is that Jánossy feels the need to introduce a second general principle. It has not been shown that the Lorentz deformation for a system is actually produced by an active boost, or change of translation motion, suffered by a system. Precisely this gap in the argument was plugged in Lorentz's original reasoning based on the theorem of corresponding states by what Janssen called Lorentz's 'generalized contraction hypothesis' (recall the discussion in section 4.4 above). This finds its expression in Jánossy's scheme as the dynamical principle: *If a connected physical system is carefully*

[27] op. cit., §183. Note that the account being given here is considerably more informal than Jánossy's.

[28] op. cit., §§275–9.

[29] In section 185, Jánossy actually states that the unobservability of the uniform translational motion of the system is a consequence of the Lorentz covariance of the laws governing the internal motion and the structure of the system. Likewise, earlier in the book (p. 64), he promises to offer Einstein-type general principles related to 'symmetry properties of the laws of physics'.

accelerated, then as a result of the acceleration it suffers a Lorentz deformation.[30] He emphasizes that this principle is 'additional' to the Lorentz principle.

It isn't, and ironically this becomes clear in Jánossy's own excellent, if brief, discussion of the process of accelerating a rigid rod in the section of the book entitled 'The mechanism of the Lorentz deformation'.[31] Jánossy analyses the boost analogue of quasi-static changes of state of a system in thermodynamics caused by slow changes of some state parameter like temperature. The second dynamical principle above turns out to follow from the Lorentz principle, once the elastic disturbances resulting from the application of external forces involved in the boost process die down and the system relaxes back to equilibrium after each step in the process. This *has* to be right, as otherwise the total dynamical description of the process, involving all possible details of the nature of the external forces and the internal constitution of the rod, would be incomplete, and truly a mysterious *coup de pouce*, to use Poincaré's term, would need to be provided by Nature. It is hard to avoid the conclusion that despite his own arguments, Jánossy somehow failed, like Lorentz before him, to understand that universal Lorentz covariance is all that is needed. And yet Jánossy's insights into the physics of the accelerative process are to be commended.

I cannot do justice here to the richness of Jánossy's 1971 study. Chapter 10 of his book extends his thinking into the problem of gravitation, and he attempts to rigorously redefine the Lorentz principle in the case of curved, dynamical spacetime. To the extent that I understand it, this analysis seems to be compatible with the position I will adopt below in Chapter 9 when it comes to understanding the local validity of SR in Einstein's theory of gravity and the physical meaning of the metric field therein.

7.5.2 J. S. Bell. Conceptual Issues

In 1976, John S. Bell (1928–1990) published a paper on 'How to teach special relativity'.[32] The paper was reprinted a decade later in his well-known book *Speakable and Unspeakable in Quantum Mechanics*, the only essay to stray significantly from the theme of the title of the book. In the paper Bell was at pains to defend a dynamical treatment of length contraction and time dilation, following 'very much the approach of H. A. Lorentz'.

Bell considered a single atom modelled by an electron circling a more massive nucleus, ignoring the back-effect of the field of the electron on the nucleus. The question he posed was: what is the prediction in Maxwell's electrodynamics (taken to be valid relative to the rest-frame of the nucleus) as to the effect on the electron orbit when the nucleus is (gently) accelerated in the plane of the orbit? Using only Maxwell's field equations, the Lorentz force law, and the relativistic formula

[30] op. cit., §193. [31] op. cit., §§195–7.

[32] Bell (1976*a*). The following analysis relies heavily on Brown and Pooley (2001).

linking the electron's momentum and its velocity, which Bell attributed to Lorentz, he concluded that the orbit undergoes the familiar longitudinal ('FitzGerald') contraction, and its period changes by the familiar ('Larmor') dilation when the motion becomes uniform. Bell went on to demonstrate that there is a system of primed variables such that the description of the uniformly moving atom with respect to them coincides with that of the stationary atom relative to the original variables, and the associated transformations of coordinates is precisely the familiar Lorentz transformation.

Bell carefully qualified the significance of this result. He stressed that the external forces involved in boosting a piece of matter must be suitably constrained in order that the usual relativistic kinematical effects such as length contraction be observed. (The force cannot be such as to disintegrate the atom!) More importantly, Bell acknowledged that Maxwell–Lorentz theory is incapable of accounting for the stability of solid matter, starting with that of the very electronic orbit in his atomic model; nor can it deal with cohesion of the nucleus. (He might also have included here the cohesion of the electron itself.) How Bell addressed this shortcoming of his model is important, and we will return to it below. Note that the positive point Bell wanted to make was about the wider nature of the 'Lorentzian' approach: that it differed from that of Einstein in 1905 in both *philosophy* and *style*.

The difference in philosophy is simply that Lorentz believed in a preferred frame of reference—the rest-frame of the ether—and Einstein did not, regarding the notion as superfluous, as we have seen. Actually, the ether plays no role at all in Bell's argument and the only justification for bringing it up at all seems to be historical. The interesting difference, rather, was that of style. Bell argues first that 'we need not accept Lorentz's philosophy to accept a Lorentzian pedagogy. Its special merit is to drive home the lesson that the laws of physics in any *one* reference frame account for all physical phenomena, including the observations of moving observers.' He went on to stress that Einstein postulates what Lorentz is attempting to prove (the relativity principle). Bell has no 'reservation whatever about the power and precision of Einstein's approach'; his point (as we mentioned in Chapter 1) is that 'the longer road [of FitzGerald, Lorentz, and Poincaré] sometimes gives more familiarity with the country'.

The point, then, is not the existence or otherwise of a preferred frame. It is how best to understand, and teach, the origins of the relativistic 'kinematical' effects. Near the end of his life, Bell reiterated the point with more insistence:

> If you are, for example, quite convinced of the second law of thermodynamics, of the increase of entropy, there are many things that you can get directly from the second law which are very difficult to get directly from a detailed study of the kinetic theory of gases, but you have no excuse for not looking at the kinetic theory of gases to see how the increase of entropy actually comes about. In the same way, although Einstein's theory of special relativity would lead you to expect the FitzGerald contraction, you are not

excused from seeing how the detailed dynamics of the system also leads to the FitzGerald contraction.[33]

There is something almost uncanny in this exhortation. Bell did not seem to be aware that just this distinction between thermodynamics and the kinetic theory of gases was foremost in Einstein's mind when he developed his fall-back strategy for the 1905 paper.

It was mentioned above that Bell drew attention in his 1976 essay to the limitations of the Maxwell–Lorentz theory in accounting for stable forms of material structure. He realized that a *complete* analysis of length contraction, say, would also require reference to forces other than of electromagnetic origin, and that the whole treatment would have to be couched in a quantum framework. In order to predict, on dynamical grounds, length contraction for moving rods and time dilation for moving clocks, Bell recognized that one need not know exactly how many distinct forces are at work, nor have access to the detailed dynamics of all of these interactions or the detailed micro-structure of individual rods and clocks. It is enough, said Bell, to assume Lorentz covariance of the complete quantum dynamics, known or otherwise, involved in the cohesion of matter.[34] Thus, once Bell recognized the limitations of his constructive pedagogical model, and stepped beyond it, his overall message was very much in the spirit of Pauli and particularly of Swann and Jánossy, though he seemed to be unaware of the arguments of the first two writers in this regard.[35]

7.5.3 Historical Niceties

For Bell, it was important to be able to demonstrate that length contraction and time dilation can be derived independently of coordinate transformations, independently of a technique involving a change of variables.[36] But as we saw in

[33] Bell (1992). Consider the analogous question in electrodynamics: why does the speed of light diminish in a transparent medium? One could say it follows from Maxwell's equations themselves in the form usually given for a linear dielectric medium of susceptibility χ. But what is the mechanism involved? One could say further that the electric field acting on a dielectric material induces an oscillating electric dipole in each molecule. The secondary radiation emitted by these dipoles combines with the primary fields to produce a single wave propagating at the reduced velocity. For some commentators, this seemingly miraculous conspiracy also calls for further analyis; for a step-by-step analysis, involving a perturbation expansion of the susceptibility χ, see James and Griffith (1992). These authors stress:

> There are no surprises here—only a comforting confirmation that the story we have told is consistent, and perhaps a somewhat deeper understanding of the mechanism by which the speed of light is reduced in a dielectric medium. (p. 309).

[34] This argument was dubbed the 'truncated Lorentzian pedagogy' in Brown and Pooley (2001).

[35] Bell cites Jánossy's 1971 book as the only modern textbook known to him taking the dynamical road of FitzGerald, Lorentz, Larmor, and Poincaré.

[36] Bell (1976*a*), p. 80.

Chapter 4, this is not strictly what Lorentz did in his treatment of moving bodies, despite Bell's claim that he followed very much Lorentz's approach.[37]

The difference between Bell's treatment and Lorentz's theorem of corresponding states that I wish to highlight is not that Lorentz never discussed accelerating systems. He didn't, but of more relevance is the point that Lorentz's treatment, to put it crudely, is (almost) mathematically the modern change-of-variables-based-on-covariance approach but interpreted in terms of what Jánossy called a Lorentz deformation and not a change of reference system. Bell's procedure for accounting for length contraction is in fact closer to FitzGerald's 1889 thinking based on the Heaviside result, summarized in section 4.3 above. In fact it is essentially a generalization of that thinking to the case of accelerating charges, followed by an application of the Lorentz force law. It is remarkable that Bell indeed starts his treatment recalling the anisotropic nature of the components of the field surrounding a *uniformly* moving charge, and pointing out that

> Insofar as microscopic electrical forces are important in the structure of matter, this systematic distortion of the field of fast particles will alter the internal equilibrium of fast moving material. Such a change of shape, the Fitzgerald contraction, was in fact postulated on empirical grounds by G. F. Fitzgerald in 1889 to explain the results of certain optical experiments.

Bell, like most commentators on FitzGerald and Lorentz, incorrectly attributes to them length contraction rather than shape deformation. But more importantly, it is unclear that Bell was aware that FitzGerald had more than 'empirical grounds' in mind, that through drawing on the work of Heaviside, he had essentially the dynamical insight Bell so nicely encapsulates. What Bell calls the 'Lorentzian pedagogy' is more aptly called the FitzGeraldian pedagogy, as was claimed in Chapter 1. But let us not exaggerate the importance of this point. In the ideas of both FitzGerald and Lorentz, and in the further articulations of Lorentz's approach by Swann and Jánossy, there is played out the lesson Bell emphasized, namely that 'the laws of physics in any *one* reference frame account for all physical phenomena, including the observations of moving observers.'

[37] op. cit., p. 77. It is noteworthy both that Bell gives no references to Lorentz's papers, and admits on p. 79 that the inspiration for the method of integrating equations of motion in a model of the sort he presented was 'perhaps' a remark of Joseph Larmor. A more recent and very careful comparative analysis of length contraction from the points of view of the electron theory and SR, and one which is more aware of the precise nature of Lorentz's thinking, is found in Dieks (1984). Dieks concludes that there is no objection in principle to the appeal for a dynamical, microscopic explanation of length contraction in SR. But he believes that relativistic theories, including general relativity, provide more scope for physical explanations of this kind. This is because, according to Dieks, general relativity explains in turn why the constructive laws such as Maxwell's equations are valid in particular frames of reference. This is correct if it is taken as part of the assumptions of general relativity that the usual form of the electromagnetic stress-energy tensor is valid. But for many commentators, the local validity of Maxwell's equations has to do with the Einstein equivalence principle (more specifically the minimal coupling principle), which is more put in by hand than explained in general relativity. We shall return to this important issue in chap. 9.

8

What is Special Relativity?

Einstein . . . wrote that in a really satisfactory theory clocks and rods should be provided self-consistently by the theory itself. Presumably, the ultimate point is the measure inherent in the *action integral* for universal dynamics. Its properties should be inherited by all partial notions and structures. In the end, it is this analysis that tells us that the geometric relations seen in physical structures are subject to physical laws.

Dierck-E. Liebscher[1]

8.1 MINKOWSKI'S GEOMETRIZATION OF SR

The dramatic words of Minkowski in 1908 have echoed down the decades: 'Henceforth space by itself, and time by itself, are doomed to fade away into mere shadows, and only a kind of union of the two will preserve an independent reality.'[2]

Today, we are so imbued with the legacy of Minkowski that Einstein's first description of his old lecturer's reformulation of SR as 'superfluous learnedness'[3] seems, at first sight, almost perverse.

Minkowski's 1908 geometrization of SR was both a crucial advance in itself (as Einstein later recognized), and a sort of stepping-stone towards Einstein's geometrical theory of gravity, or general relativity. The 1908 paper altered the feel of the theory. It introduced the space-time diagram and terms like 'world-line', front and back light 'cones' and 'proper time', that are still popular today. Interestingly, the piece of nomenclature that Minkowski introduced with greatest emphasis, that of the 'postulate of the absolute world', or the 'world-postulate', never stuck.

This was the postulate that 'natural phenomena', and by this Minkowski clearly means the fundamental laws of physics,[4] have as their covariance group G_c, whose elements are coordinate transformations which preserve the expression

[1] Liebscher (2005), §10.2.

[2] Minkowski (1909). An elegant modern treatment of the geometry of Minkowski space-time is found in Liebscher (2005).

[3] Pais (1982), p. 152.

[4] See particularly the penultimate para. in section I.

$c^2 \mathrm{d}t^2 - \mathrm{d}x^2 - \mathrm{d}y^2 - \mathrm{d}z^2$ (what we would call today the isometries of the Minkowski metric).[5] The laws of physics are Lorentz covariant. But by expressing this notion in 4-dimensional geometrical language, and showing how it applies even to central tenets of the Lorentz–Larmor theory of the electron, Minkowski felt he had shown how 'the validity of the world-postulate without exception ... now lies open in the full light of day'. Minkowski's aim was that of clarification and illumination.

As with Einstein's 1905 paper, Minkowksi's paper contained distinct kinematical and dynamical arguments. I shall comment briefly on each in turn.

8.1.1 Kinematics

Minkowski starts with the 'fundamental axiom' that (in modern parlance) a particle can always be regarded as at rest relative to some inertial coordinate system, which would not be possible were the speed of light not a maximal speed. (Note that Minkowski is right to introduce this as an extra postulate—it does not follow straightforwardly from the world-postulate, nor of course from Einstein's 1905 postulates.) He then goes on to claim that the existence of rigid bodies is only consistent with mechanics whose covariance group is the (Galilean) group G_∞, since a theory of optics compatible with G_c and rigid bodies would lead to a violation of the relativity principle. Note that by 'rigid' Minkowski means a body whose dimensions are unaltered by motion, an infelicitous, though common, choice of terminology. But the important point here is that Minkowski goes on to construct what is clearly a crucial argument for him in favour of his 4-dimensional picture of space-time.

Minkowski first claims that the FitzGerald–Lorentz 'contraction' hypothesis (which he attributes solely to Lorentz) 'sounds extremely fantastical, for the contraction is not to be looked upon as a consequence of resistances in the ether, or anything of that kind, but simply as a gift from above,—as an accompanying circumstance of the circumstance of motion.'[6]

Minkowski, like Poincaré before him, seems quite oblivious to the dynamical plausibility arguments that both Lorentz and FitzGerald (not to mention Larmor) gave for the motion-induced change in dimensions in rigid bodies. He now claims that 'the Lorentzian contraction hypothesis is completely equivalent to the new conception of space and time, which indeed makes the hypothesis much more intelligible.'

Minkowski proceeds to give the graphic argument which has been repeated so often in the relativistic literature: that (in modern vernacular) the world tube of a rigid body is sliced into space-like sections in different ways by observers in

[5] By Lorentz covariance I shall mean covariance with respect to the 10-dimensional inhomogeneous Lorentz group, or the 'Poincaré' group, which incorporates rigid rotations and translations as well as boosts.

[6] op. cit., part II.

relative motion. This claim to intelligibility is so important, and came to be so influential, that we must dwell on it in detail.

The first point that must be made is that Minkowski is not claiming that it is space-time structure in and of itself that renders the phenomenon of length contraction intelligible. Actually, Minkowski seems nervous about the physicality of empty space.[7] Much more importantly, *Minkowski's paper is an attempt to spell out the mathematical consequences, only partially understood by Einstein, of the world-postulate, and in particular of the claim that the laws of physics are Lorentz covariant.*[8] There is in the paper an explicit analogy with the nature of Euclidean 3-space and its association with the fact that the laws of classical mechanics are invariant under rigid spatial rotations, which similarly preserve the distance between points. It is the world-postulate that accounts for the unity of Minkowski space-time, for the fact that there exists an invariant 4-dimensional metric.[9] At the very end of the 1908 paper, Minkowski refers to a pre-established harmony between pure mathematics and physics, and I take the former to be the Minkowski geometry of pure space-time and the latter to be the Minkowski geometry of natural phenomena exemplified by the laws of physics which adhere to the world-postulate.

Thus the world-postulate is responsible for the phenomenon of length contraction, but Minkowski is presenting an account of this connection that appears to be different from that of Pauli, Swann, and Bell. Minkowski is reading length contraction straight off the induced geometry.

The reading is curious. It is odd to intimate that only the geometric argument confers intelligibility to the FitzGerald–Lorentz contraction, when the geometry itself is a direct consequence of a structural feature of the dynamics of non-gravitational interactions. The dynamical line of argument outlined in the last chapter is not only intelligible, *it explains what the Minkowski geometry means physically*. It accounts for the fact that ideal rods and clocks, in different states of uniform motion, and irrespective of their constitution, exemplify the physics of coordinate transformations, of the group G_c, and thus survey the Minkowski geometry. It is beautiful to see how simply the inverse of the Lorentz factor γ can be distilled out of the Minkowski metric in the way that Minkowski first did it and which has been repeated on innumerable occasions since, but the fact that this

[7] When Minkowski comes, at the end of part I, to explicate the meaning of the universal constant c that appears in the group G_c, he first states that it is the velocity of light in empty space. Then he writes: 'To avoid speaking either of space or emptiness, we may define this magnitude in another way, as the ratio of the electromagnetic to the electrostatic unit of electricity.'

[8] Minkowski's historical reasoning at the end of part II is odd. He misleadingly claims that Lorentz gave 'local' time a 'physical construction . . ., for the better understanding of the hypothesis of contraction'. He then assigns credit for the first recognition of the idea that 't and t' are to be treated identically' to Einstein, with no mention of Poincaré. But then, most curiously, he states that neither Einstein nor Lorentz 'made any attack on the problem of space', even though a 'violation of the concept of space' is 'indispensable for the true understanding of the group G_c'.

[9] This crucial feature of Minkowski's logic has not gone entirely unnoticed. Jon Dorling stressed it in private communication with Michel Janssen; see Janssen (1995), p. 267.

is the length-change factor of actual physical rigid bodies needs supplementary arguments of the kind developed in the previous chapter.

8.1.2 Dynamics

Electrodynamics gave Minkowski the real opportunity to show the simplifying power of the 4-dimensional approach. His reconstruction in the 1908 paper of the Liénard–Wiechert potentials associated with an arbitrarily moving charge is based on the geometry (and in particular the notion of orthogonality) of 4-vectors, and his treatment of the action of one moving charge on another relies specifically on his analysis of the velocity and acceleration 4-vectors and their orthogonality. Minkowski writes of this latter treatment

> When we compare this statement with previous formulations ... of the same elementary law of the ponderomotive action of moving point charges on one another, we are compelled to admit that it is only in four dimensions that the relations here taken under consideration reveal their inner being in full simplicity, and that on a three-dimensional space forced upon us a priori they cast only a very complicated projection.[10]

The elegance and simplifying power of the 4-dimensional approach to relativistic dynamics is probably Minkowski's greatest legacy. In particular, where Oliver Heaviside had vastly simplified the formulation of Maxwell's equations by using the 3-vector formalism which demonstrates manifest covariance relative to the symmetries of Euclidean space, so Minkowski produced an equally important simplification of electrodynamics based on the 4-vector formalism, which analogously secures manifest covariance with respect to the symmetries of Minkowski space-time. That, at any rate, is how Einstein in the last years of his life described Minkowski's legacy.

> Minkowksi's important contribution to the theory [SR] lies in the following: Before Minkowski's investigation it was necessary to carry out a Lorentz-transformation on a law in order to test its invariance under such transformations; he, on the other hand, succeeded in introducing a formalism such that the mathematical form of the law itself guarantees its invariance under Lorentz-transformations. By creating a four-dimensional tensor-calculus he achieved the same thing for the four-dimensional space which the ordinary vector-calculus achieves for the three spatial dimensions. He also showed that the Lorentz transformation (apart from a different algebraic sign due to the special character of time) is nothing but a rotation of the coordinate system in the four-dimensional space.[11]

It is often forgotten that it was Poincaré, in 1905, who first understood the 'rotational' nature of the Lorentz transformations, and provided the first insights into the 4-vector calculus. It is very curious that Minkowski never cites Poincaré's great 1906 paper on the dynamics of the electron, or even mentions his name in the 1908 paper.

[10] op. cit., section IV. [11] Einstein (1969), p. 59.

8.2 MINKOWSKI SPACE-TIME: THE CART OR THE HORSE?

There is a long-standing tradition, at least within the philosophy literature, to go beyond Minkowski himself, or rather to invert his logic, and to consider Minkowski space-time as the basis of an explanatory account of the Lorentz covariance of physical laws (indeed of the world-postulate), and hence relativistic kinematics. In Michael Friedman's *Foundations of Space-Time Theories*, for example, special relativity is the theory of a space-time manifold equipped with a flat Minkowski metric.[12] It is particularly clear in the section of Friedman's book on the reality of space-time structure[13] that even before the manifold is endowed with the Maxwell or any other dynamical field, the Minkowski geometry is understood as physically real and primordial.

The reader will recall from the last chapter that Einstein admitted that constructive theories like the kinetic theory of gases provide more insight into the nature of things than principle theories like thermodynamics and special relativity. *The question naturally arises as to whether there is a constructive version of SR.*[14] The approach advocated by Friedman is tantamount to the claim that the Minkowski geometry provides the structure that is needed in a constructive account of the theory.

In discussing Poincaré's preference for the Lorentz–FitzGerald approach to length contraction over Einstein's, Friedman wrote more recently:

> [T]he crucial difference between the two theories, of course, is that the Lorentz contraction, in the former theory, is viewed as a result of the (electromagnetic) forces responsible for the microstructure of matter in the context of Lorentz's theory of the electron, whereas the same contraction, in Einstein's theory, is viewed as a direct reflection, independent of all hypotheses concerning microstructure and its dynamics, of a new kinematical structure for space and time involving essential relativized notions of duration, length, and simultaneity. In terms of Poincaré's hierarchical conception of the sciences, then, Poincaré locates the Lorentz contraction (and the Lorentz group more generally) at the level of experimental physics, while keeping Newtonian structure at the next higher level (what Poincaré calls mechanics) completely intact. Einstein, by contrast, locates the Lorentz contraction (and the Lorentz group more generally) at precisely this next higher level, while postponing to the future all further discussions of the physical forces and material structure actually responsible for the physical phenomenon of rigidity. The Lorentz contraction, in Einstein's hands, now receives a direct ***kinematical*** interpretation.[15]

The talk of a preference for one theory over the other might suggest that we are dealing with two incompatible viewpoints. On one side one has a truly

[12] Friedman (1983), chap. IV. [13] op. cit. §4, chap. VI.

[14] This question was raised in Brown (1993), Brown (1997), and Balashov and Janssen (2003); the answers provided were essentially positive, but were not in agreement as to what the constructive account is, as we shall see. [15] Friedman (2002), pp. 211–12.

constructive space-time interpretation of SR, involving the postulation of the structure of Minkowski space-time as an ontologically autonomous element in the models of the phenomena in question. In this picture, length contraction is to be given a constructive explanation in terms of Minkowski space-time because the behaviour of complex material bodies is constrained somehow to 'directly reflect' its structure, in a way that is 'independent of all hypotheses concerning microstructure and its dynamics'. If one were to adopt such a viewpoint there would seem little room left for the opposing viewpoint, according to which the explanation of length contraction is ultimately to be sought in terms of the dynamics of the microstructure of the contracting rod.

In fact, it is not clear that Friedman has these two opposing pictures in mind. Although he claims that Poincaré keeps Newtonian structure at the level of 'mechanics', if one is committed to the idea that Lorentz contraction is the result of a structural property of the forces responsible for the microstructure of matter—as is defended in this book—then one should believe that Minkowskian, rather than Newtonian, structure is the appropriate kinematics for mechanics. The appropriate structure is Minkowski geometry *precisely because* the laws of physics of the non-gravitational interactions are Lorentz covariant. Equally one can postpone (as Einstein did) the detailed investigation into the forces and structures actually responsible for the phenomena in question without thereby relinquishing the idea that these forces and structures are, indeed, actually responsible for the phenomena, and, hence, for space-time having the structure that it has.

Precisely the opposite idea has recently been defended by Balashov and Janssen, who wrote:

> [D]oes the Minkowskian nature of space-time explain why the forces holding a rod together are Lorentz invariant or the other way around? Our intuition is that the geometrical structure of space(-time) is the *explanans* here and the invariance of the forces the *explanandum*. To switch things around, our intuition tells us, is putting the cart before the horse.[16]

It has to be noted that Balashov and Janssen's target is a particular 'neo-Lorentzian' interpretation of SR, in which space-time structure is supposed to be Newtonian and in which there is supposed to be a preferred frame, a position very different from anything defended in this book. But what Balashov and Janssen find unattractive in this interpretation also applies to the dynamical treatment of relativistic kinematics outlined in the previous chapter, and even to Minkowski's own position:

> In the neo-Lorentzian interpretation it is, in the final analysis, an unexplained coincidence that the laws effectively governing different sorts of matter all share the property of Lorentz invariance, which originally appeared to be nothing but a peculiarity of the laws governing electromagnetic fields. In the space-time interpretation this coincidence is explained by

[16] Balashov and Janssen (2003).

tracing the Lorentz covariance of all these different laws to a common origin: the space-time structure posited in this interpretation.

Here we are at the heart of the matter. It is wholly unclear how this geometrical explanation is supposed to work.[17] To help clarify matters, let us look at other situations in physics where absolute geometric structure makes an appearance.

8.2.1 The Cases of Configuration and 'Kinematic' Space

It has been recognized since the middle of the nineteenth century, particularly through the work of Jacobi, that the evolution of a system of N particles in Newtonian mechanics can be described by way of a geodesic (maximal straightness) principle involving paths in the configuration space of the system, this space having a curved Riemannian geometry.[18]

Another case of geometric reasoning that is closer to home concerns relativistic velocity (or 'kinematic') space.[19] It was remarked earlier that the Fresnel drag coefficient, so important in explaining the absence of first order effects in early ether-wind experiments, was recognized in 1907 by Max von Laue to be a consequence of the relativistic rule for transforming velocities. In 1908, Arnold Sommerfeld found a geometrical account of this rule: the original velocity of the body in question $\mathbf{u}$, the velocity of the moving frame $\mathbf{v}$, and their 'resultant' (the transformed velocity $\mathbf{u}'$) form the sides of a spherical triangle on a sphere of imaginary radius ic. A year later, V. Varičak pointed out that 3-dimensional velocity space in SR has constant negative curvature with radius c, i.e. it is a Lobachevskian geometry. (The corresponding space in Galilean–Newtonian kinematics is flat.) A velocity and its 'Lorentz' transform form a pair of points in this space, and their relation is obtained by solving for Lobachevskian triangles; the geometry of such triangles is equivalent to to that of Sommerfeld's spherical triangles of imaginary radius. (In 1921, Pauli pointed out a new interpretation of the conformal structure of Minkowski space-time in terms of the 3-dimensional Lobachevksian geometry.)[20]

A striking feature of this Lobachevksian geometry is that the non-commutativity of the transformation rule, in the general case where $\mathbf{u}$ and $\mathbf{v}$ do not lie in the same direction, is an immediate consequence of the curvature of kinematic space. And in 1913, Emil Borel considered the holonomy associated with parallel-transport of a vector around a closed path in the space, predicting that

[17] A detailed critique of the position defended by Friedman, and by Balashov and Janssen, is found in Brown and Pooley (2004). In this chapter, some points are taken from this critique and some new ones are added.

[18] See Lanczos (1970). I am grateful to Jeremy Butterfield for stressing the relevance of this early use of non-Euclidean absolute geometry in physics.

[19] In the following treatment, I rely heavily on section 4.2.6 of Stachel (1995), to which the reader is referred for further references. See also Liebscher (2005), Appendix B.

[20] Pauli (1981), p. 74, fn. 111.

a torque-free angular momentum vector undergoes a precession. This is precisely the Thomas precession effect, famously rediscovered in 1926.

Borel's work in 1913 demonstrates the suggestive power of geometric reasoning in physics. But the issue that concerns us is that of explanation. Do we want to say that the non-commutativity of velocity transformations in SR, and the Thomas precession are *caused*, or *explained by* the existence of curvature in relativistic velocity space? Do we likewise want to say that the curvature of the configuration space is causing the motion of the N-body system in mechanics to be what it is? Note a crucial difference between these cases and general relativity: the geometry here is not a dynamical agent, there are no non-trivial equations of motion which couple it with matter. It is absolute.

8.2.2 The Projective Hilbert Space

Following the seminal work of Michael Berry in 1984 on systems undergoing cyclic adiabatic evolution, it has come to be realized that there exists an important geometrical structure in the quantum formalism related to the phase of a system undergoing standard Schrödinger evolution. Consider a system undergoing cyclic (but not necessarily abiabatic) evolution, so that during the temporal interval $[0, T]$, the system's final and initial states coincide up to a phase factor: $|\psi(T)\rangle = \exp(i\phi)|\psi(0)\rangle$, where ϕ is an arbitrary real number. When projected on to ray space, i.e. the projective Hilbert space $\mathcal{P}$, this evolution defines a closed path. Now suppose we have the idea of subtracting from the total phase ϕ the accumulation of local phase changes produced by the motion on this path. By a 'local phase change' is meant the quantity $\delta\phi\,(\psi_t, \psi_{t+\delta t}) = -i\langle\psi(t)|\mathrm{d}/\mathrm{d}t|\psi(t)\rangle\delta t$ (choosing units in which $\hbar = 1$). We are subtracting then from the total phase the quantity

$$\phi_d \equiv -i \int_0^T \langle\psi(t)|\mathrm{d}/\mathrm{d}t|\psi(t)\rangle dt = -\int_0^T \langle\psi(t)|H|\psi(t)\rangle \mathrm{d}t, \qquad (8.1)$$

where H is the Hamiltonian responsible for the cyclic motion. Because it depends on the Hamiltonian, the quantity ϕ_d is called the *dynamical phase*. Now what we are left with after the subtraction, $\phi_g \equiv \phi - \phi_d$, is the *geometric phase*, formulated by Yakir Aharonov and the late Jeeva Anandan in 1987. It is reparametrization invariant, i.e. independent of the speed at which the path in $\mathcal{P}$ is traversed. Moreover, it takes the same value for all the (infinity of) evolutions in the Hilbert space which project onto the given closed path in $\mathcal{P}$. It is natural then to interpret it as the holonomy associated with 'parallel transport'—transport in which there is no local phase change—around the closed curve in $\mathcal{P}$. The existence of the geometric phase testifies to the existence of an absolute, non-flat connection, or curvature, in $\mathcal{P}$. It turns out this is the curvature associated with the so-called Study–Fubini metric on $\mathcal{P}$.[21]

[21] For a review, see Anandan (1992).

The recognition that there is a feature of Schrödinger evolution that is indifferent to the dynamical details specified by the Hamiltonian, or at least the choice of Hamiltonian within an infinite relevant class, and that depends on only the fixed path in $\mathcal{P}$, has led to a significant geometrical reformulation of quantum mechanics. In particular, the symplectic structure of $\mathcal{P}$, and its role as a metric space have been clarified. The situation is clearly reminiscent of the geometric reformulation of SR by Minkowski. But, again, it seems unnatural to say that the phenomenon of geometric phase is *caused* by the geometry of the projective Hilbert space of a quantum system. Is not this geometric structure merely an elegant codification of the existence of a universal, or Hamiltonian-independent, feature of the Schrödinger dynamics of cyclic motions, albeit one that remained hidden for many years?[22] But then what grounds are there for assigning a causal role to Minkowski space-time?

8.2.3 Carathéodory: The Minkowski of Thermodynamics

It is of considerable relevance to our concerns that a development in thermodynamics itself took place in 1909 that is strikingly analogous to that instigated by Minkowski in relation to SR in 1908. In 1909, Constantin Carathéodory (1873–1950), a Berliner of Greek origin and who had as it happens studied under Minkowski when writing his doctoral thesis in mathematics a few years earlier, published a new formulation of thermodynamics.[23] This formulation was far more abstract than that found in the well-known writings of Clausius, Kelvin, and Planck on the subject. Carathéodory wanted to turn attention away from cyclic processes of the type Sadi Carnot had discussed, and in particular away from the delicate notion of heat, in favour of the structure of the space of equilibrium states available to a thermodynamic system. He did to this space something akin to what Minkowski had done to space-time.

Carathéodory assumed that the thermodynamic phase space Γ of equilibrium states is an N-dimensional differentiable manifold equipped with the usual Euclidean topology. Global coordinate systems are provided by the values of N thermodynamic parameters, some (actually one in the case of 'simple' systems) 'thermal' and the rest 'deformation' variables. Different coordinate systems of this kind are available and no one is privileged. Instead of postulating a metric on the space, Carathéodory introduced the relation of *adiabatic accessibility* between pairs of points, clearly analogous to the causal connectibility relation in the conformal geometry of Minkowski space-time, and famously postulated that in any neighbourhood of any point p in Γ, there exists at least one point q that is not adiabatically accessible from p.

[22] It is noteworthy that Anandan, one of the pioneers of the geometric approach to quantum mechanics (see Anandan (1991)), wrote a careful defence of the dynamical underpinning of absolute space-time geometry (see Anandan (1980)).

[23] Carathéodory (1909).

On the basis of this postulate, which replaces the traditional second law, and a number of crucial continuity assumptions, Carathéodory was able to show that for simple systems a foliation of Γ exists, a division of the space into a continuum of non-intersecting $(N-1)$–dimensional surfaces, such that any continuous curve confined to a given surface represents a quasi-static reversible process between end-points of the curve involving continuous change in the deformation coordinates. More importantly, each surface could be identified by the value of a continuous real parameter S, such that for any two points p and q not on the same surface, q being adiabatically accessible from p means $S(q) > S(p)$. Finally, Carathéodory was able to prove that S and T play the familiar roles of entropy and absolute temperature respectively.

Actually, this account needs to be qualified. What Carathéodory in fact was able to deduce from his assumptions is that for any two points p and q not on the same surface, i.e. not connected by a reversible quasi-static path, when q is adiabatically accessible from p then *either* $S(q) > S(p)$ *or* $S(q) < S(p)$. In other words, irreversible processes either all involve an increase in entropy or all involve a decrease in entropy. Carathéodory accepted that this is a weaker condition than the analogous conclusion of strict entropy increase that follows from, say, Kelvin's formulation of the second law which rules out perpetual motion machines of the second kind, and in which the notion of heat is explicitly referred to. Carathéodory considered it to be a matter of *experience* whether in his scheme the monotonic law of entropy change is one of increase or decrease.[24]

The 1909 formulation proved to be the inspiration for a number of subsequent approaches to thermodynamics based likewise not on the notion of heat and Carnot-type cycles, and the efficiency of heat engines, but rather on properties of the thermodynamic phase space Γ. The latest significant development in this tradition is the 1999 axiomatization due to Lieb and Yngvason, in which the differentiable nature of Γ is no longer assumed, but a variant of the adiabatic accessibility relation retained, now explicitly a preorder on the space.[25] This approach avoids a number of criticisms that were levelled at Carathéodory's 1909 formulation.[26] Be that as it may, the work of Carathéodory is clearly analogous to that of Minkowski in 1907. In both cases, an existing theory is recast in a more abstract, geometrical mould.

Is it reasonable to say that Carathéodory transformed thermodynamics from the paradigmatic principle theory to a constructive theory? Of course there is no categorical answer to this question, for the simple reason that the distinction, as has

[24] It is clear that this appeal to experience only works if a background arrow of time exists, relative to which the change of entropy can be compared. Carathéodory was silent about what this pre-thermodynamic arrow of time is, but in this reticence he was not alone. The founders of thermodynamics were happy to discuss (quasi-static) cyclic motion for a heat engine, which requires saying which of two nearby equilibrium states is the earliest, but little was said as to what this meant in practice.

[25] Lieb and Yngvason (1999).

[26] See Uffink (2001).

already been stressed, is not categorical. Nonetheless there is certainly the appearance in the very postulates of Carathéodory's scheme of structure that is defined in theoretical terms rather than observational terms. It is less phenomenological than the traditional approach. But the new scheme still seems a long way short of the insight provided by the mixture of combinatorial and dynamical arguments that is characteristic of statistical mechanics. This point has been aptly expressed by Peter Landsberg, whose remarks were intended to apply to the formal axiomatic approach just as much as to the traditional formulation of thermodynamics.

> Since the nature of the thermodynamical variables and their number can vary within wide limits, the basic theoretical framework of thermodynamics must be kept very general. This has the advantage of giving the theory a wide range of application, but this is balanced by the drawback that thermodynamic reasoning is in general unsuitable for giving insight into the details of physical processes.
>
> This last observation, together with the remark that thermodynamics leaves microscopic variables on one side, leads to the conclusion that a thermodynamic theory is necessarily incomplete. For any system to which thermodynamics can be applied, a more exhaustive theory should exist which yields insight into the detailed physical processes involved. This deeper theory must also give some account of the nature of equilibrium states, and therefore of fluctuation phenomena, and of course it must lead to the thermodynamics of the system. Such theories are provided by statistical mechanics.[27]

I do not know if Einstein took any interest in Carathéodory's work in thermodynamics, but there is little doubt that despite its importance as a template in 1905, Einstein's long-term attitude towards thermodynamics was that of a non-fundamental, perhaps one could say effective, theory. He would consistently endorse the Boltzmannian combinatorial definition as the most fundamental rendition of entropy, and referred at the end of his life to the concepts of classical thermodynamics as 'untenable in the long run'.[28]

Nor did Einstein ever appear to have been tempted to say that Minkowski provided the elements of a constructive formulation of SR. Indeed, there is a hint that geometrical notions are not fundamental in the theory in Einstein's *Autobiographical Notes*, as we saw in section 8.1. I think it is likely that Einstein's reasons are closely analogous to the sentiments expressed by Landsberg concerning

[27] Landsberg (1990), p. 4.

[28] In one of his last letters, in which he touched on the widely accepted role of classical concepts in quantum mechanics, Einstein wrote:

> I believe however that the renunciation of the objective description of 'reality' is based upon the fact that one operates with fundamental concepts which are untenable in the long run (like f.i. classical thermodynamics).

This 1955 letter to Andre Lamouche is cited in Stachel (1986), p. 376. Nonetheless, as Stachel has pointed out, the chances of success of a fully constructive approach to fundamental physics at times seemed remote to Einstein, particularly as a result of his failure to erect a unified field theory. Stachel notes (p. 360) in some of Einstein's letters in the 1950s that touch on his scientific method, the advocacy of a kind of blurring of the principle vs. constructive theory distinction.

thermodynamics, including Carathéodory's formulation. Einstein would have expected a constructive theory of SR to do much more than anything Minkowski had done in his 1908 paper. Minkowski had implemented a geometrical programme initiated by Poincaré in 1906, but this was a far cry from the new theory of the interaction of matter and radiation that Poincaré was hoping would supersede Lorentz's theory of the electron and rigorously, systematically account for length contraction or rigid rods, the deformability of the electron, etc. These ambitious demands on a truly constructive explanation of relativistic kinematics were shared by both Einstein and Poincaré, but of the two only Einstein was content in the meantime to make do with a weaker principle theory explanation.

8.3 WHAT DOES ABSOLUTE GEOMETRY EXPLAIN?

It is doubtful at best whether the geometries of the configuration space in classical mechanics, the 'kinematic' space in SR, the projective Hilbert space in quantum mechanics, or the space of equilibrium states in thermodynamics, play the kind of explanatory role that the space-time interpretation of SR attributes to Minkowski geometry. Why should space-time geometry be any different? It might be thought that space and time are somehow more fundamental physically than the other spaces, or more accessible to the senses, or that they combine to form the arena of physical events. In short, that they are more real. But is not this reasoning question-begging?[29] The reader may recall that in section 3.2.5 above, it was mentioned that Einstein for a limited period of time (after 1905) sought a geometric explanation of inertia. The position defended by Balashov and Janssen (and many others) is an attempt to extend Einstein's logic in this sense to an explanation of the Lorentz covariance of physical laws. In order to assess further the plausibility of this line of thinking, let's see how far Einstein was right about inertia.

[29] A defence of the view that the conformal structure of space-time is a structural property of Maxwellian electrodynamics, and that even the Euclidean nature of space is determined by the dielectric properties of the medium in electrostatics, is found in Sternberg (1978) and Guillemin and Sternberg (1984). (Thanks go to Gordon Belot for bringing this work to my attention.) Here is how the point is made in relation to space in both works:

> The statement that it is the dielectric properties of the vacuum that determines Euclidean geometry is not merely a mathematical sophistry. In fact, the forces between charged bodies in any medium are determined by the dielectric properties of that medium. Since the forces that bind together macroscopic bodies as we know them are principally electrostatic in nature, it is the dielectric property of the vacuum which fixes our rigid bodies. We use rigid bodies as measuring rods to determine the geometry of space. It is in this very real sense that the dielectric properties of the vacuum determine Euclidean geometry.

The related notion that the Euclidean geometry of space is a consequence of the Euclidean symmetries of the fundamental laws of non-gravitational physics is defended in Brown and Pooley (2004).

8.3.1 The Space-time 'Explanation' of Inertia

It was a source of satisfaction for Einstein that in developing the general theory of relativity (GR) he was able to eradicate what he saw as an embarrassing defect of SR: violation of the *action–reaction principle*. Leibniz held that a defining attribute of substances was their both acting and being acted upon. It would appear that Einstein shared this view. He wrote in 1924 that each physical object 'influences and in general is influenced in turn by others'.[30] It is 'contrary to the mode of scientific thinking', he wrote earlier in 1922, 'to conceive of a thing . . . which acts itself, but which cannot be acted upon.'[31] But according to Einstein the space-time continuum, in both Newtonian mechanics and special relativity, is such a thing. In these theories space-time upholds only half of the bargain: it acts on material bodies and/or fields, but is in no way influenced by them.

It is important to be clear about what kind of action Einstein thought is involved here. Although he did not describe them in these terms, it seems that he had in mind the roles of the four-dimensional absolute affine connection in each case, as well as that of the conformal structure in SR. The connection determines which paths are geodesics, or straight, and hence determines the possible trajectories of force-free bodies. The null cones in SR in turn constrain the possible propagation of light. These structures form what Einstein would come to call, with tongue in cheek, the ether.

> The inertia-producing property of this ether [Newtonian space-time], in accordance with classical mechanics, is precisely *not* to be influenced, either by the configuration of matter, or by anything else. For this reason, one may call it 'absolute'. That something real has to be conceived as the cause for the preference of an inertial system over a noninertial system is a fact that physicists have only come to understand in recent years . . . Also, following the special theory of relativity, the ether was absolute, because its influence on inertia and light propagation was thought to be independent of physical influences of any kind . . . The ether of the general theory of relativity differs from that of classical mechanics or the special theory of relativity respectively, insofar as it it is not 'absolute', but is determined in its locally variable properties by ponderable matter.[32]

The success in salvaging the action–reaction principle was not confined in GR to the fact that the space-time metric field (which of course determines both the connection, by the principle of metric compatibility, and the conformal structure) is dynamical, that it is a solution of Einstein's field equations. In the early 1920s, when he wrote the above comment, Einstein had still not discovered an important aspect of his theory of gravitation: the fact that the field equations themselves underpin the geodesic principle. This principle states that the world-lines

[30] Einstein (1924).

[31] Einstein (1922, 55–6). For a recent discussion of the action–reaction principle in modern physics, see Anandan and Brown (1995).

[32] Einstein (1924).

of force-free test particles are constrained to lie on geodesics of the connection. It is important for our purposes to dwell briefly on the significance of this fact.

In 1924, Einstein thought that the inertial property of matter (to be precise, the fact that particles with non-zero mass satisfy Newton's first law of motion, not that they possess such inertial mass) requires explanation in terms of the action of a real entity on the particles. It is the space-time connection that plays this role: the affine geodesics form ruts or grooves in space-time that guide the free particles along their way. In GR, on the other hand, this view is at best redundant, at worst problematic, something that was still not widely appreciated in the early 1920s. For it follows from the form of Einstein's field equations that the covariant divergence of the stress-energy tensor field $T_{\mu\nu}$, that object which incorporates the 'matter' degrees of freedom, vanishes.

$$T^{\mu}{}_{\nu;\mu} = 0. \tag{8.2}$$

This result is about as close as anything is in GR to the statement of a conservation principle, and it came to be recognized as the basis of a proof, or proofs, that the world-lines of a suitably modelled force-free test particles are geodesics.[33] The fact that these proofs vary considerably in detail need not detain us. The first salient point is that the geodesic principle for free particles is no longer a postulate but a theorem. GR is the first in the long line of dynamical theories, based on that profound Aristotelian distinction between natural and forced motions of bodies, that *explains* inertial motion. The second point is that the derivations of the geodesic principle in GR also demonstrate its limited validity. In particular, it is not enough that the test particle be force-free. It has long been recognized that spinning bodies for which tidal gravitational forces act on its elementary pieces deviate from geodesic behaviour. What this fact should clarify, if indeed clarification is needed, is that it is not simply *in the nature* of force-free bodies to move in a fashion consistent with the geodesic principle.[34] It is not an essential property of localized bodies that they run along the ruts of space-time determined by the affine connection, when no other dynamical influences are at play. In Newtonian mechanics and SR, the conspiracy of inertia is a postulate, and its explanation by way of the affine connection is no explanation at all.

And it is here that Einstein and Nerlich part company with Leibniz, and even Newton. For both Leibniz *and Newton*, absolute space-time structure is not the sort of thing that acts at all. If this is correct, and it seems to be, then *neither Newtonian mechanics nor SR represent*, pace *Einstein, a violation of the action–reaction principle*, because the space-time structures in both cases are neither acting nor being acted upon. Indeed, we go further and agree with Leibniz that they are not real entities in their own right at all.

[33] We return to this issue in chap. 9.

[34] A rare recognition of this point in the philosophical literature is found in Sklar (1977) and Sklar (1985), chap. 3.

It is well known that Leibniz rejected the reality of absolute Newtonian space and time principally on the grounds that their existence would clash with his principles of Sufficient Reason and the Identity of Indiscernibles. Nonentities do not act, so for Leibniz space and time can play no role in explaining the mystery of inertia. Newton seems to have agreed with this conclusion, but for radically different reasons, as expressed in his pre-*Principia* manuscript *De Gravitatione*. For Newton, the existence of absolute space and time has to do with providing a structure, necessarily distinct from ponderable bodies and their relations, with respect to which it is possible systematically to define the basic *kinematical* properties of the motion of such bodies. For Newton, space and time are not substances in the sense that they can act, but are real things nonetheless.[35] It is now known, however, that the job can be done without postulating any background space-time scaffolding, and that at least a significant subset, perhaps *the* significant subset, of solutions to any Newtonian theory can be recovered in the process.[36]

Recall Nerlich's remark reported in section 2.2.5 to the effect that force-free particles have no antennae, and that they are unaware of the existence of other particles. That *is* the prima facie mystery of inertia in pre-GR theories: how do all the free particles in the world know how to behave in a mutually coordinated way such that their motion appears extremely simple from the point of view of a family of privileged frames? To appeal, however, to the action of a background space-time connection in which the particles are immersed—to what Weyl called the 'guiding field'—is arguably to enhance the mystery, not to remove it. There is no *dynamical coupling* of the connection with matter in the usual sense of the term.

As remarked earlier, it is of course non-trivial that inertia can be given a *geometrical* description, and this is connected with the fact that the behaviour of force-free bodies is universal: it does not depend on their constitution. But, again, what is at issue is the arrow of explanation. It is more natural in theories such as Newtonian mechanics or SR to consider the 4-connection as a codification of certain key aspects of the behaviour of particles and fields.[37] One faces a similar choice in parity-violating theories: do orientation fields play an explanatory role

[35] It is worth stressing that its lack of causal influence is Newton's sole reason for refraining from calling space a substance. It is therefore at least misleading to deny that Newton (1962) was a substantivalist.

[36] See Barbour and Bertotti (1982) and Barbour (1999). For discussion see Belot (2000) and Pooley and Brown (2002).

[37] Robert DiSalle put this point well in 1995:

> Space-time theories are not the sort of theory Einstein thought they were, because they don't really make the sort of metaphysical claim that he thought they make—in particular, space-time theories do not claim that some unobservable thing is the cause of observable effects. Instead, they make a more restricted, but perhaps more profound and certainly more useful claim: that particular physical processes, governed by established physical laws, can be represented by aspects of geometrical structure in the universe. (DiSalle (1995).)

in such theories, or are they simply codifications of the coordinated asymmetries exhibited by the solutions of such theories?[38]

8.3.2 Mystery of Mysteries

In the dynamical approach to length contraction and time dilation that was outlined in the previous chapter, the Lorentz covariance of all the fundamental laws of physics is an unexplained brute fact. This, in and of itself, does not count against the approach: all explanation must stop somewhere. What is required if the so-called space-time interpretation is to win out over this dynamical approach is that it offer a genuine explanation of universal Lorentz covariance. This is what is disputed. Talk of Lorentz covariance 'reflecting the structure of space-time posited by the theory' and of 'tracing the invariance to a common origin' needs to be fleshed out if we are to be given a genuine explanation here, something akin to the explanation of inertia in general relativity. Otherwise we simply have yet another analogue of Molière's dormative virtue. (We shall see later that nothing in GR causes us to rescind this view.)

In fact space-time theorists often accommodate, if grudgingly, theories in which the symmetries of space-time structure are *not* reflected in the symmetries of the laws governing matter. They do not question the coherence of such theories (as they should); rather they seek to rule them out on the grounds of explanatory deficiencies when compared to their preferred theory.[39] This shows that, as matter of logic alone, if one postulates space-time structure as a self-standing, autonomous element in one's theory, it need have no constraining role on the form of the laws governing the rest of content of the theory's models.[40] So how is its influence on these laws supposed to work?[41] How in turn are rods and clocks supposed to know which space-time they are immersed in? This mystery becomes even more acute when it is borne in mind that there is a growing number of 'bimetric' theories in the literature: attempts to modify Einstein's general relativity theory which involve two metric fields, with different contributions to the gravitational dynamics. A brief account of one such theory will be given in the next chapter. The question that must be faced is how rods, clocks, free particles, and light rays come to survey at most one of them.

38 See Pooley (2003, 272–4).

39 For a clear indication of this, see the discussion in Earman (1989), pp. 46–7.

40 See in this connection Brown (1993).

41 A more sustained discussion of how Minkowski space-time provides a putative common origin for the 'unexplained coincidence' in Lorentz's theory that both matter and fields are governed by Lorentz covariant laws, is to be found in Janssen's detailed recent analysis of the differences between the Einstein and Lorentz programmes in Janssen (2002*b*). It is also covered in Janssen's wider investigation of 'common origin inferences' in the history of science; see Janssen (2002*a*). The claim that neither of these papers succeeds in clarifying how space-time structure can act as a 'common origin' of otherwise unexplained coincidences is defended in Brown and Pooley (2004).

8.4 WHAT IS SPECIAL RELATIVITY?

It will probably not have escaped the attention of readers that the range of matters relativistic treated so far has been narrow. Relativistic kinematics have dominated the discussion. There is of course much more to SR, and arguably the most important part has not been treated. In the fifth of his 1905 papers, Einstein reverted to a theme that had already made its appearance in the literature on the ether theories: the connection between mass and electromagnetic energy in special situations. In formulating his famous equation $E = mc^2$, Einstein was the first to postulate the proportionality between mass and all forms of (rest) energy. This is no more a direct consequence of the Lorentz transformations than were Einstein's original definitions of relativistic force and momentum. Indeed, the latter were to give way (with Einstein's approval) to simpler definitions provided by Planck, which in turn were incorporated by Minkowski into the four-momentum vector.

But in restricting ourselves to just a part of the edifice of SR, we can still ask: What precisely did Einstein offer? One present-day authority on the history and foundations of relativity theory, John Stachel, considers what the consensus view might plausibly have been in the community of experts had Einstein made no appearance.

> The work of Lorentz, Poincaré, and others suggests that, without Einstein's contribution, the consensus version might not have made a clear distinction between kinematic and dynamic effects, but interpreted such things as length contraction, time dilation and increase of mass with velocity as dynamical effects, caused by motion relative to the ether frame.... Emphasis would then have been placed on factors leading to the undetectability of absolute velocity, rather than on the complete equivalence of all inertial frames.[42]

I think this view encapsulates the difficulty most relativists have today in accepting what has been defended throughout this book: the true *lack* of a clear distinction between kinematic and dynamic effects (in particular in the context of length contraction and time dilation). The difficulty arises because the valid dynamical insights provided by the trailblazers like FitzGerald, Lorentz, and Poincaré are contaminated in most physicists' minds by association with the philosophy of ether. One of the chief aims of Chapter 4 was to demonstrate that the ether played no truly essential role in the pre-1905 thinking that led to the discovery of length contraction and time dilation (to the limited extent that the latter was appreciated by Larmor and Lorentz). All the real work was being done by Maxwell's equations and the idea that molecular forces inside chunks of matter might mimic electromagnetic forces in a specific sense (a point that as we saw in Chapter 7 was stressed particularly by Bell). In the cases of Lorentz and Poincaré, it is clear that the role of the ether was essentially that of a peg to hang the electromagnetic field on,

[42] Stachel (1995), p. 272.

and even then Poincaré was at times doubtful. Giving up the ether-as-substratum prejudice simply does not entail rejecting the dynamical basis of the trailblazers' arguments or their modern variants. The difference that Stachel regards as critical between the undetectability of absolute velocity and the complete equivalence of all inertial frames is, in this sense, chimerical.

In conclusion, Einstein's contribution was not to establish a clear-cut divide between kinematics and dynamics, though both the organization and terminology of his 1905 paper were misleading in this respect. We saw in Chapter 7 that Einstein was aware of this defect in his initial formulation of SR. What his contribution amounted to, rather, was the demonstration (a) of the full operational significance of the Lorentz transformations, and (b) that the latter could be obtained by imposing simple phenomenological constraints on the nature of the fundamental interactions in physics. But the importance of this second feature is to some extent epoch-dependent: in 1905 something like this derivation was all that was available given the theoretical maelstrom in physics and in particular the advent of the quantum with all its subversive, or revolutionary, potential. But now?

8.4.1 The Big Principle

Stachel has argued that Einstein, having been 'mesmerized' by the problem of the nature of light,[43] failed to see that the light postulate and the relativity principle are qualitatively different and should not be mixed together. Stachel correctly emphasized that the relativity principle is 'universal in scope'[44] whereas the light postulate is taken (as we have seen in Chapter 5) from the ether theory of electrodynamical phenomena. 'Such traces of its electrodynamical origins exposed the new kinematics to avoidable misunderstandings and attacks over the years.'[45]

For Stachel, Einstein overlooked the importance of the 1910 work of Ignatowski, who 'showed that the special theory does not require the light postulate'.[46]

An earlier commentator who emphasized this point is Jean-Marc Lévy-Leblond. In 1976 he wrote 'the lesson to be drawn from more than half a century is that

[43] Stachel (1995), p. 270.

[44] Stachel actually claims that the relativity principle is kinematical in nature, whereas in this book the relativity principle has been described as dynamical. I argued in Chapter 5 that in Einstein's hands it was not essentially different from the principle defended by Newton—a constraint on the nature of the fundamental interactions. Indeed, the apparently curious fact that such a 'dynamical' principle could be used to constrain 'pure' kinematics in the form of coordinate transformations was examined in detail in both chap. 2 (in the context of the Keinstein's derivation of the Galilean transformations) and chap. 5 (in the context of Einstein's derivation of the Lorentz transformations). But it seems that this particular difference with Stachel is merely terminological. He simply defines kinematical to be synonymous with principle theory; see footnote [4], p. 323.

[45] Stachel (1995), p. 270. [46] op. cit., pp. 272, 277.

special relativity up to now seems to rule *all* classes of phenomena, whether they depend on electromagnetic, weak, [or] strong . . . interactions.'[47]

Lévy-Leblond emphasized that the Lorentz covariance of the laws of physics does not entail the existence of zero-mass bodies, or even that the photon mass is exactly zero. This is an important point, and we return to it in the next chapter. Like Stachel, Lévy-Leblond extolled the virtues of the Ignatowski derivation; indeed, in the paper under consideration he rediscovered it!

A warning was sounded earlier (section 6.4.1) as to whether the Ignatowski transformations are indeed relativistic in nature. They are clearly weaker than the Lorentz transformations, in the sense that they rule out less. In particular, they do not exclude the Galilean transformations, and hence are less falsifiable (in the Popperian sense) than the Lorentz transformations. Unless the magnitude of the invariant speed is established, the Ignatowski group can hardly be equated with the Lorentz group. In short, an empirical element (that for the purposes of Stachel and Lévy-Leblond should not refer to light in an essential way) over and above Ignatowski's postulates is needed.

Be that as it may, we must not lose sight of the distinction between the content of a theory and its mode of discovery, or even its axiomatic formulation. If it is the content of SR that we are after, then as Lévy-Leblond himself attested

Relativity theory, in fact, is but the statement that all laws of physics are invariant under the Poincaré group (inhomogeneous Lorentz group). The requirement of invariance, when applied to a classfication of the possible fundamental particles . . . permits but does not require the existence of zero-mass objects . . .[48]

This is the real point, and in making it, Lévy-Leblond was doing nothing more than reiterating Einstein's own view. As Stachel has stressed,[49] Einstein came to regret the emphasis in his 1905 paper on the behaviour of light. Here is how he put it in 1935.

The special theory of relativity grew out of the Maxwell electromagnetic equations. But it came about that even in the derivation of the mechanical concepts and their relations the considerations of those of the electromagnetic field played an essential role. The question as to the independence of these relations is a natural one because the Lorentz transformation, the real basis of special-relativity theory, in itself has nothing to do with the Maxwell theory.[50]

As we saw in section 6.7, Einstein emphasized again in his 1949 *Autobiographical Notes* that the central principle of SR is the Lorentz covariance of all the fundamental laws of (non-gravitational) physics, and somewhat confusingly tried to tie this big principle into the thermodynamical, or principle-theory, methodology

[47] Lévy-LeBlond (1976), p. 271. [48] ibid. [49] Stachel (1995), pp. 9, 271–2.
[50] Einstein (1935).

he used in 1905.[51] At any rate, the nature of the big principle—what Minkowski called the world postulate—has often led commentators to remark that SR has the character of a 'meta-theory', or that it has a 'transcendental' flavour. It doesn't tell you how things interact; rather it tells you how any theory of interaction should behave. The big principle is a *super law* whose function, like that of all symmetry principles, is in Eugene Wigner's words . . . 'to provide a structure or coherence to the laws of nature, just as the laws of nature provide a structure and coherence to the set of events'.[52]

At its most fundamental, SR is a theory that lies somewhere between a pure principle theory (like thermodynamics, or Einstein's 1905 version of SR) and a fully constructive theory (like statistical mechanics). The big principle is a universal constraint on the nature of the non-gravitational interactions. It is a 'restricting principle', as Einstein himself put it, and does not determine the exact form of the dynamics in question. It is this facet of the theory that so bewildered Poincaré, who was already well aware of the big principle, and even attributed it to Lorentz. For Poincaré the principle was the germ of a theory, not a theory in its own right. It must have been galling for him to see Einstein's SR given such prominence, when a good part of what the young man seemed to be doing was merely to postulate what he and Lorentz had been trying to prove the hard way.

8.4.2 Quantum Theory

The one and only meeting between Einstein and the great French polymath took place at the Solvay Conference in Brussels in 1911, a year before Poincaré's death. Maurice de Broglie later recalled: 'I remember one day at Brussels, while Einstein was explaining his ideas, Poincaré asked him, "what mechanics are you using in your reasoning?" Einstein answered: "No mechanics" which appeared to surprise his interlocutor.'[53]

Einstein may have been discussing his hypothesis of the light quantum rather than SR, but as both were based, in different ways, on thermodynamic considerations, it may not much matter for our purposes. For his part, Einstein found Poincaré's incomprehension frustrating,[54] and from a modern point of view Poincaré stance might well seem short-sighted, if not high-handed.

Criticisms can, I think, be justly made of Poincaré's position. It does not appear that he really appreciated the full significance of Einstein's operational analysis of the Lorentz coordinate transformations associated with boosts. Poincaré continued to regard length contraction as a separate assumption, and never seems to have appreciated the existence of time dilation. Most importantly, as we saw in Chapter 4, Poincaré never understood that the dynamical origins of length

[51] Recall that the relativity principle that appears as an axiom in both the 1905 Einstein and the 1910 Ignatowski derivations must not under pain of circularity contain reference to the precise form of the inertial coordinate transformations.

[52] Wigner (1964).

[53] Galison (2004), p. 297.

[54] ibid.

contraction could be tied up with the big principle of Lorentz covariance. However, there is an aspect of his thinking that deserves commendation, at Einstein's expense. Einstein was justified in 1911 to reply 'no mechanics', because as he repeated throughout his life, no successful constructive theory of the constitution of matter and radiation existed in the early years of SR. Under these circumstances, the self-confessed 'sin' of treating rods and clocks as primitive entities within SR was forgivable. But within two decades or so a new mechanics did emerge: quantum mechanics. As I have stressed earlier, by the time of Einstein's death exactly fifty years after his *annus mirabilis*, much insight had been gained within quantum theory as to the detailed nature of material structure. And as Swann emphasized in the 1940s (recall the discussion of his arguments in the previous chapter), an answer of sorts had appeared to the obvious question related to the big principle: Lorentz covariance of what? Einstein chose largely to ignore this development, at least in the context of his original 'sin'. The irony is that what Poincaré wanted he did not live to see; what Einstein lived to see, and partially instigated, he did not want.[55]

An analogy with the biological sciences suggests itself. Charles Darwin's theory of evolution by natural selection relies critically on two empirically based claims: the existence of variation, or differences between individual members of a species, and heredity, or the ability of these members to pass on the differences to succeeding generations. Darwin himself realized that a biological mechanism must exist to bolster these claims, but never succeeded in specifying what it was. He developed a simple, but amazingly fertile theory of how slow adaptation to the changing environment by plants and animals accounts for the immense variety of species seen today in the natural world. But the analogy with a principle theory in physics, and in particular Einstein's 1905 theory of relativity is obvious: Darwin's theory of natural selection in its original form presupposed the existence of a microscopic mechanism whose nature was obscure, but whose existence was required by a critical combination of fact and theory. The obscurity of the mechanism did not prevent early success for the theory within the community of life scientists. But it was of course the 'new synthesis' based on subsequent knowledge of genetics, chromosomes, and DNA that filled the gap, and re-established the primacy of Darwin's theory after the doldrums it was to go through around the turn of the twentieth century.

[55] It is in the nature of the argument that Swann, and to a lesser extent Bell, gave that the appeal to quantum theory involves nothing other than the Lorentz covariance of its laws. The argument is obviously far from fully constructive. To provide a quantum theoretical analogue of Bell's constructive pedagogical model, particularly one involving inter-atomic forces, would be an ambitious undertaking, outlandishly so if the starting point is the most fundamental theory we have: quantum field theory. I am unaware of a systematic derivation of the stable structure of even the hydrogen atom within relativistic quantum field theory. And even today, relativistic treatments of the cohesion of matter based on quantum mechanical models of interaction are not thick on the ground. One starting point for the physics of interatomic forces in molecules and crystals is based on relativistic Projected-Dirac-Breit Hamiltonians within the framework of the Born–Oppenheimer approximation; see Schwartz *et al.* (1991).

It is dangerous to push this analogy too far. Amongst other things, there is every reason to think Darwin himself would have been thrilled by the genetic revolution had he lived to see it. It is intriguing that this revolution not merely provided the missing mechanism in natural selection, but changed the philosophy of biology. I have in mind the relatively recent insight, due principally to William Hamilton and developed and popularized by Richard Dawkins, that the agent of evolution is neither the group nor the individual but the gene. This notion has had a profound effect on the understanding of such key issues as altruism and sexual reproduction in evolution, not to mention the little issue of our own significance in the cosmic order.[56] There is no way that appreciation of the role quantum theory plays in the most fundamental account of the relativistic behaviour of rods and clocks can lead to as profound a change in the philosophy of relativity theory, and it is a pity all of this has nothing to do with sex. But it is part of the full picture, and it paves the way to some of the arguments developed in the next chapter.

[56] For a recent, highly readable account of these developments, see Sykes (2004).

9

The View from General Relativity

Space tells matter how to move. Matter tells space how to curve.

Charles W. Misner, Kip S. Thorne, and
John Archibald Wheeler[1]

9.1 INTRODUCTION

In 1954, a year before his death, Albert Einstein wrote in a letter to Georg Jaffe

> You consider the transition to special relativity as the most essential thought of relativity, not the transition to general relativity. I consider the reverse to be correct. I see the most essential thing in the overcoming of the inertial system, a thing which acts upon all processes, but undergoes no reaction. The concept is in principle no better than that of the center of the universe in Aristotelian physics.[2]

As we saw in section 8.3.1, Einstein viewed the 'inertial system', the privileged coordinate system associated with the rigid geometric structure of Minkowski space-time, as explaining the principle of inertial motion for force-free bodies in SR. Whether Einstein was right about space-time geometry in SR 'acting' on bodies, and whether therefore in the theory there is a violation of the action–reaction principle, may be questioned (and we did question it above), but there remains the fundamental difference between SR and the general theory of relativity (GR): in GR the metric tensor (and hence the affine connection, given their compatibility) becomes a dynamical agent. It undeniably acts and is acted upon. But in what sense is it *geometry*, or *space-time structure* itself that acts?

In his 1923 book *The Mathematical Theory of Relativity*, Arthur Eddington distinguished between two chains of reasoning in GR. The first familiar one starts with the existence of the four-dimensional space-time interval ds, whose meaning is the usual one associated with the readings of physical rods and clocks and possibly light rays. From the metric field $g_{\mu\nu}$ associated with the interval

$$ds^2 = g_{\mu\nu}dx^{\mu}dx^{\nu} \tag{9.1}$$

[1] Misner *et al.* (1973), p. 5 [2] See Stachel (1986), p. 377.

(recall we are using the Einstein summation convention for repeated indices) the Einstein tensor is built up, which finally is equated with the stress-energy tensor related to matter. The other less familiar chain of reasoning 'binds the physical manifestations of the energy tensor and the interval; it passes from matter as now defined by the energy-tensor to the interval regarded as the result of measurements made with this matter.'[3]

The aim of this chapter is to take up the challenge of outlining this second chain of reasoning, at least as I see it. In developing this reasoning, it will be argued that the dynamical underpinning of relativistic kinematics that has been defended in this book is consistent with the structure and logic of GR.[4] It will be seen as we go along that chronogeometrical significance of the $g_{\mu\nu}$ field is not an intrinsic feature of gravitational dynamics, but earns its spurs by way of the strong equivalence principle.

The first part of the chapter concerns various aspects of the nature of Einstein's field equations. The second part concerns the operational significance of the metric field. Appendix A deals with Einstein's changing views on the significance of general covariance in GR.

9.2 THE FIELD EQUATIONS

9.2.1 The Lovelock–Grigore Theorems

A 1969 theorem due to D. Lovelock[5] establishes that gravitational actions (in the sense of the principle of 'least' action) are severely constrained by the requirement of invariance under arbitrary coordinate transformations:

$$x^{\mu} \longrightarrow x'^{\mu} = x^{\mu} + \epsilon \xi^{\mu} \tag{9.2}$$

where ξ^{μ} is an arbitrary vector field and ϵ is small. Specifically, Lovelock proved that in a space-time of four or fewer dimensions, any strictly invariant action that is second-order in the metric field $g^{\mu\nu}$ (i.e. depends on at most second derivatives of the metric with respect to the coordinates) and that gives rise to second-order field equations, must be associated with a Lagrangian density $\mathcal{L}_{\text{grav}}$ which is a linear combination of the Hilbert Lagrangian density and a cosmological term:

$$S_{\text{grav}} = \int_{\Omega} \mathcal{L}_{\text{grav}} \, \mathrm{d}^4x = \int_{\Omega} (aR + \lambda)\sqrt{|g|} \, \mathrm{d}^4x, \tag{9.3}$$

[3] Eddington (1965), p. 146.

[4] The arguments in this chapter are a development of the logic given in Brown and Pooley (2001), in which the possibility of extending the dynamical approach advocated by Bell into GR was defended.

[5] Lovelock (1969).

where R is the curvature scalar and g is the determinant of $g^{\mu\nu}$. Note that $|g|$ appears in the square root rather than the usual $-g$, because no assumptions about the signature of $g^{\mu\nu}$ are made in the theorem.[6]

It is easy to overestimate the role that invariance is playing in this theorem. For one thing, it is not at all trivial that a second-order action be required to produce only second-order Euler–Lagrange equations, and not fourth-order equations as might be expected (i.e. twice the order reached in the Lagrangian). In fact the reason that the usual Hilbert Lagrangian density $R\sqrt{-g}$ achieves this feat can best be seen by recalling how Einstein introduced his version of the Lagrangian formulation of general relativistic dynamics.

Einstein was originally interested in unimodular coordinates satisfying the condition $\sqrt{-g} = 1$, as he thought that such a restriction produced a substantial simplification of gravitational dynamics. (We will return to this remarkable aspect of Einstein's thinking, and how it was abandoned by late 1916, in Appendix A.) For these coordinates he introduced the Lagrangian density

$$\mathcal{L}_{\text{grav}} = g^{\mu\nu}\Gamma^{\alpha}{}_{\mu\beta}\Gamma^{\beta}{}_{\nu\alpha}\sqrt{-g}, \tag{9.4}$$

where the $\Gamma^{\alpha}{}_{\beta\gamma}$ represent the (metric compatible) connection coefficients.[7] For arbitrary coordinates, Einstein's Lagrangian density—often called the $\Gamma-\Gamma$ Lagrangian—became[8]

$$\mathcal{L}_{\text{grav}} = g^{\mu\nu}\left(\Gamma^{\alpha}{}_{\mu\beta}\Gamma^{\beta}{}_{\nu\alpha} - \Gamma^{\beta}{}_{\mu\nu}\Gamma^{\alpha}{}_{\alpha\beta}\right)\sqrt{-g}. \tag{9.5}$$

Now both the Hilbert action (the action in (9.3) with $\lambda = 0$) and the action obtained from the $\Gamma-\Gamma$ Lagrangian (9.5) give rise, under Hamilton's principle

[6] A word on conventions and notation. The Lorentzian signature of the metric $g^{\mu\nu}$ is $(+---)$. Greek indices run from 0 to 3; $x^0 = ct$, t being coordinate time. For any geometrical object F, $F_{,\mu} = \partial_\mu F = \partial F/\partial x^\mu$ and $F_{;\mu} = \nabla_\mu F$.

[7] In 1915, Einstein was unaware of the geometrical meaning of the Christoffel symbol, a function of the gravitational potential $g_{\mu\nu}$ and its first derivatives, as a connection, or rule for defining parallel transport. Indeed, such an insight would have to wait for work by Hermann Weyl, principally. Before this development, Einstein referred to his gravitational theory as an exercise in the theory of general invariants; it was not a fully 'geometrical' theory yet in his mind.(See Stachel (1995), p. 292.) However that may be (and it is good not to lose sight of the fact that the notion of 'geometry' is very elastic) it is noteworthy that the motivation for Einstein's choice of the Lagrangian density (9.4) is analogous with the form of the Lagrangian density for the free Maxwell field, $-(1/4)F_{\mu\nu}F^{\mu\nu}\sqrt{-g}$ (see section 9.4 below). A major breakthrough in Einstein's long and tortuous road of discovery was the realization in late 1915 that the components of the gravitational field $\Gamma^{\alpha}{}_{\mu\beta}$ are not simply the first derivatives of the gravitational potential, or more specifically $(1/2)g^{\alpha\lambda}g_{\lambda\mu;\beta}$ as in his 1913 'Entwurf' theory, but are defined as in the Christoffel symbols. This point has been emphasized in the impressive recent historical study of Einstein's discovery of the field equations by Janssen and Renn (2005), which also highlights the importance of the role of the action principle in Einstein's thinking from 1913 on. A very nice, briefer historical account is found in Sauer (2004).

[8] The Lagrangian density (9.4) appeared in Einstein (1915, 1916*a*); the Lagrangian density (9.5) in Einstein (1916*b*). Note that Einstein's connection coefficients were the negative of the usual Christoffel symbols.

applied to arbitrary variations in $g^{\mu\nu}$, to the same Euler–Lagrange equations,[9] which are Einstein's famous matter-free or vacuum field equations

$$G_{\mu\nu} \equiv R_{\mu\nu} - \frac{1}{2} R g_{\mu\nu} = 0, \tag{9.6}$$

where $R_{\mu\nu}$ is the Ricci tensor. (These equations are equivalent to $R_{\mu\nu} = 0$.) That second-order equations are obtained from the $\Gamma - \Gamma$ Lagrangian density which is first-order is not surprising. But it can be shown that if one subtracts the $\Gamma - \Gamma$ density from the Hilbert Lagrangian density the result is a second-order term that takes the form of a *total divergence* of a first-order functional of $g_{\mu\nu}$.[10] And it is well known that two Lagrangian densities that differ only by a total divergence term give rise to the same Euler–Lagrange equations under Hamilton's principle.

Einstein's original approach to the Lagrangian formulation of the field equations reminds us of a point that is sometimes overlooked in texts on GR: that the gravitational action integral need not be a scalar invariant as in (9.3).[11] Indeed, a strengthened version of the Lovelock theorem was reported by D. Grigore in 1992,[12] concerning the class of first-order gravitational actions that are quasi-invariant, i.e. invariant up to a total divergence. This new theorem, the proof of which is daunting, purports to establish that, *whatever* the dimensionality of space-time, the Lagrangian density appearing in this action must take the form of a linear combination of the $\Gamma - \Gamma$ Lagrangian (9.5) and a cosmological term. As in the Lovelock theorem, the result does not depend on the signature of the metric being Lorentzian.

Further words of caution are in order when the results of Lovelock and Grigore are taken to show the importance of general covariance as a constraint on gravitational dynamics. General covariance is a symmetry of the equations of motion, such as the field equations (9.6) above. It is trivially satisfied by any equations that can be written in the tensor calculus. But conceptually there is a distinction between dynamical symmetries and what are sometimes called variational symmetries. The former are transformations of the dependent and possibly independent variables that take solutions of the equations of motion into solutions. The latter are transformations of these variables which render the action invariant or quasi-invariant (i.e. invariant up to a surface term, which because of Gauss's theorem corresponds to the Lagrangian density being invariant up to a divergence). A sufficient condition for (quasi-) invariance of the action to imply the

[9] The reader will recall that this is the principle requiring the first-order variation in the action to vanish under arbitrary infinitesimal variations in the (inverse of the) metric, as well as its first derivatives in the case of the Hilbert action, which themselves vanish on the boundary of the region of integration in the action integral.

[10] For details see, for instance, Eddington (1965), §58, Schrödinger (1985), pp. 100–1, and Dirac (1966), §26.

[11] Cf., e.g., Misner *et al.*, p. 503. The issue is not a new one. The standard action for a free particle in Newtonian mechanics is not invariant under a pure Galilean boost, but it is 'quasi-invariant' (see below). For further discussion, see Brown and Brading (2002).

[12] Grigore (1992).

corresponding dynamical symmetry with respect to a given transformations of variables is that all the dependent variables in the action are subject to Hamilton's principle, or in other words, that all the fields are dynamical.[13] In fact, Lovelock and Grigore assumed something even stronger: that the only field that represents the gravitational potential is $g_{\mu\nu}$, that no other geometrical object fields, dynamical or otherwise, appear in the action. The 1961 Brans–Dicke theory of gravitation also proposes second-order equations in $g_{\mu\nu}$, and has an invariant second-order action distinct from Hilbert's (up to a cosmological term). That the theory is not a counter-example to the Lovelock theorem is due to the fact that the (matter-free) Brans–Dicke action depends on a scalar field as well as $g^{\mu\nu}$.

The demand that nothing other than the $g_{\mu\nu}$ field is needed to describe the dynamics of pure gravity has occasionally been referred to in the mainstream literature as the 'the principle of general covariance',[14] but it clearly goes much further than the mere requirement of coordinate generality, or the requirement that the field equations be written in the tensor calculus. We shall discuss the nature and justification of this weaker requirement in Appendix A, but it is useful now to look briefly at one of its important consequences.

9.2.2 The Threat of Underdetermination

Suppose we have a system of partial differential equations involving one or more fields (which may be just the components of a single tensor field), and a spacelike hypersurface defined by $x^0 = \text{const}$. We may wish to know whether, given suitable 'Cauchy' data on the hypersurface involving the fields and their partial x^0-derivatives, the differential equations uniquely determine the fields in the neighbourhood of any point on the hypersurface. This is the case when both an analytic solution of the equations exists at this point, and it is possible to reconstruct uniquely the whole sequence of partial x^0-derivatives of each of the fields there. For then we can obtain the coefficients of the power series expansion of each of the fields about the point in question, using the well-known connection between the coefficients and such partial derivatives.

Now it is straightforward to see that any system of equations that is generally covariant cannot satisfy this requirement, since the equations will necessarily be formally *underdetermined*. This follows from the simple fact that given a solution of the equations, a distinct solution can always be constructed by a coordinate transformation which reduces to the identity transformation in a four-dimensional neighbourhood of the Cauchy, or 'initial value' hypersurface, but which is arbitrary elsewhere. (Recall that the coordinate transformation (9.2) depends on the arbitrary vector field ξ^μ, and therefore can depend on space and time.) Mathematically distinct solutions are thus consistent with the same initial data.

[13] For a recent discussion, see Brown and Holland (2004) and the references therein.
[14] See, e.g., Wald (1984), p. 57.

An important theorem due to Cauchy and Kovalevskaya establishes a sufficient condition for the existence of unique solutions to the Cauchy problem. The condition is that the differential equations in question take the 'normal form': in our case they are such as to allow for the highest partial x^0-derivative appearing in the equations to be expressed for each of the fields as an analytic function of the fields and their derivatives, providing no higher derivative with respect to coordinates other than x^0 appears. This permits the calculation of the whole sequence of partial x^0-derivatives mentioned above and ensures the convergence of the power series expansion.[15] It can be shown generally that for as many fields as there are arbitrary functions defined in the symmetry (in our case four), the Cauchy–Kovalevskaya condition fails.[16]

In the present case of matter-free GR it is relatively easy to see how this works. The (twice contracted) Bianchi identity states that the covariant divergence of the Einstein tensor $G_{\mu\nu}$ vanishes independently of the validity of the field equations:

$$G^{\mu}{}_{\nu;\mu} = 0. \tag{9.7}$$

It is easily shown that these identities can be rewritten in the form

$$\left(\sqrt{-g}G^{\mu}{}_{\nu}\right)_{,\mu} - \Gamma^{\mu}{}_{\nu\sigma}\sqrt{-g}G^{\sigma}{}_{\mu} = 0. \tag{9.8}$$

Inspection of (9.8) indicates that $G^{0}{}_{\mu}$ contains no second derivatives of the metric with respect to x^0, because the second term contains at most second derivatives. Now given Einstein's field equations (9.6), it follows that four of these, namely $G_{0\mu} = 0$ do not take the 'normal' form mentioned earlier in the context of the Cauchy problem, and are 'constraints' (recall we are taking the Cauchy hypersurface to be spacelike); they express conditions between components of the metric tensor and their x^0-derivatives on the hypersurface. Only six of the ten Einstein vacuum field equations thus propagate the Cauchy data. These same equations propagate the constraints themselves, in the sense that it is a consequence of the field equations that the x^0-derivatives of the constraints vanish.[17] Note however that the six equations involving the G_{ij} $(i, j = 1, 2, 3)$ *can* be arranged in the normal form, given that x^0 is time-like.

Formal underdetermination of this kind may seem at first sight disastrous for attempts to model a deterministic world. Indeed, Einstein was reluctantly led to question the viability of general covariance in his new metric theory of gravity between 1913 and 1915 on precisely these grounds, until he realized that mathematically distinct solutions of covariant field equations generated by coordinate

[15] Discussion of the Cauchy–Kovalevskaya theorem can be found in John (1978), pp. 76–86, and Dennery and Krzywicki (1967), p. 333.

[16] A useful general treatment of this issue is found in Anderson (1967), §4–6, 4–7.

[17] See, e.g., Bergmann and Komar (1980) and Schmidt (1996).

transformations (so-called active diffeomorphisms) are *empirically* equivalent.[18] This equivalence turns out to be a general feature of theories with 'local' symmetries: no observable differences exist between different solutions of the dynamical equations that are related by local, or 'gauge' transformations, and so the threat to predictability is lifted.

But the philosophical implications of the diffeomorphism invariance of GR are still being debated, and a large literature exists today related to the question whether the notion of space-time as a substance is viable in the light of this symmetry. One thing seems obvious: if the bare, differential space-time manifold is a real entity, then different solutions of Einstein's field equations that are related by diffeomorphisms correspond to different physical states of affairs.[19] The theory is incapable of predicting which of the different possible worlds is realized, but all of them are, as we have seen, empirically indistinguishable. The simplest (and to my mind the best) conclusion, and one which tallies with our usual intuitions concerning the gauge freedom in electrodynamics, is that the space-time manifold is a non-entity. In this case the different, diffeomorphically related worlds are not only observationally indistinguishable, they are one and the same thing.[20]

9.2.3 Matter

All these considerations apply equally well in the case where 'matter' is present[21] and represented by the stress-energy tensor $T_{\mu\nu}$. In the action principle approach, the total action S now contains a matter-related term:

$$S = S_{\text{grav}} + S_{\text{matter}} = \int_{\Omega} \left(\mathcal{L}_{\text{grav}} + \mathcal{L}_{\text{matter}}\right) \mathrm{d}^4 x. \tag{9.9}$$

It is normally assumed that the $\mathcal{L}_{\text{matter}}$ depends not just on the variables associated with the matter fields but also on $g_{\mu\nu}$ but not its derivatives. The assumption that the total action can be broken up into two components, one of which depends only on $g_{\mu\nu}$, is non-trivial, as we see below.

Application of Hamilton's principle of the stationarity of the action under variations of the inverse of the metric $g^{\mu\nu}$ yields the familiar result that the the sum of the variational (or Hamiltonian) derivatives of $\mathcal{L}_{\text{grav}}$ and $\mathcal{L}_{\text{matter}}$ vanishes:

$$\frac{\delta \mathcal{L}_{\text{grav}}}{\delta g^{\mu\nu}} + \frac{\delta \mathcal{L}_{\text{matter}}}{\delta g^{\mu\nu}} = 0. \tag{9.10}$$

[18] For treatments of Einstein's struggle with the 'hole problem', see Norton (1988) and Stachel (1989).

[19] But this has been contested; see Maudlin (1990).

[20] As we noted in section 2.1 above, this was apparently Einstein's own conclusion. However, for a defence of what is now called in the philosophy literature 'sophisticated' space-time substantivalism, see Pooley (2006).

[21] The word 'matter' in GR usually means all those fields and particles whose degrees of freedom are distinct from gravitational degrees of freedom. In particle physics, there is a subdivision of reality into 'matter' (leptons and quarks) and 'radiation' (fields like the electromagnetic, gravitational, and gluon fields). See Rovelli (1997), note 10, p. 219.

Defining finally the symmetric second rank tensor

$$T_{\mu\nu} \equiv \frac{-2}{\sqrt{-g}} \frac{\delta \mathcal{L}_{\text{matter}}}{\delta g^{\mu\nu}}, \tag{9.11}$$

and suppressing the cosmological constant so $\mathcal{L}_{\text{grav}} = \kappa^{-1} R\sqrt{-g}$, say ($\kappa = 8\pi G$, where G is Newton's gravitational constant; the similarly weighted Einstein $\Gamma - \Gamma$ Lagrangian density would do just as well) we arrive at the general field equations

$$G_{\mu\nu} \equiv R_{\mu\nu} - \frac{1}{2} g_{\mu\nu} R = 8\pi G T_{\mu\nu}. \tag{9.12}$$

There follow some remarks about these famous equations.

Meaning of the field equations

The equations are, unlike the field equations in electromagnetism, non-linear; and they are very hard to solve in any systematic fashion. The equations can be re-arranged to take the following form, where the roles of R, the Ricci scalar, and T (recalling that $T \equiv T^{\mu}{}_{\mu}$) are reversed:

$$R_{\mu\nu} = 8\pi G \left(T_{\mu\nu} - \frac{1}{2} g_{\mu\nu} T \right). \tag{9.13}$$

These equations will be true, remarkably, if that for any one component holds in all local inertial coordinate systems everywhere in space-time. In such coordinates $g_{\mu\nu} = \text{diag}(-1, 1, 1, 1)$ (see below), so $g_{00} = -1$, and

$$T = -T_{00} + T_{11} + T_{22} + T_{33}, \tag{9.14}$$

so the field equations are equivalent to the fact that

$$R_{00} = \frac{8\pi G}{2} (T_{00} + T_{11} + T_{22} + T_{33}) \tag{9.15}$$

holds in every local inertial frame. Now T_{00}, the flow of momentum in the t-direction, is just the energy density ρ, and $T_{11} \equiv T_{xx}$, the flow of momentum in the x-direction, is the momentum in that direction P_x, etc.

The operational significance of $R_{00} \equiv R_{tt}$ in an inertial frame of reference can be demonstrated as follows. Imagine a small ball of free-falling test particles all of which are assumed to have world-lines that are geodesics of the metric, with four-velocities U^{μ}. (The justification of this geodesic assumption will be taken up shortly.) The covariant divergence of U^{μ}, or 'expansion' $\theta = U^{\mu}{}_{;\mu}$, is a measure of the rate of change of the three-volume of the ball. This will depend on initial conditions, and we shall assume that it is initially zero. But gravity will determine the rate of change of the expansion, or the acceleration of the volume. When the particles are all initially at rest,[22] so that $U^\mu = (1, 0, 0, 0)$, it can be shown that

$$\frac{d\theta}{dt} = -R_{\mu\nu}U^\mu U^\nu = -R_{tt}. \tag{9.16}$$

Thus we end up in these special circumstances with the simple result

$$\frac{d\theta}{dt} = \frac{-8\pi G}{2}(\rho + P_x + P_y + P_z). \tag{9.17}$$

The beauty of this formulation of the field equations is the ease with which it reveals some qualitative properties of Einsteinian gravity. First, suppose a massive body is situated at the centre of the ball of test particles; its positive mass density is a form of energy density, and because of the minus sign in (9.17) the volume of the ball will shrink. Gravity attracts. Not a surprising result, but comforting. Second, gravity depends on pressure—a feature that has no Newtonian analogue—and acts in the opposite sense to the normal, 'direct' action of pressure which is expansive. (Normally, the gravitational effect of pressure is swamped by that of energy density, but in neutron stars, for instance, the two effects are comparable. On the human scale, all terms on the RHS of (9.17) are small: gravity is a weak force.) Finally, consider the case where matter is absent, or $T_{\mu\nu} = 0$. Since the rate of change of volume is initially zero, and it now follows from (9.17) that the acceleration is too, it must be the case that volume remains constant, even though curvature can distort the ball. Tidal effects will occur in a ball in free fall at the surface of the earth. Even though we are describing the physics relative to a local inertial coordinate system, and hence the tidal effects will be negligible, they nonetheless occur in such a way that the vertical elongation and horizontal squeezing are consistent with zero acceleration in volume. The same thing happens when gravitational waves pass through the ball.[23]

A noteworthy feature of the matter-free case is this. It follows from the reasoning in section 9.2.2 above that R_{00} contains no second x^0-derivatives of $g_{\mu\nu}$, and yet the second-time derivative of the volume of the ball of test particles is determined by it. It turns out that the same feature obtains in the general case where matter is present, as long as the pressure terms in the RHS of (9.17) vanish, since analogous reasoning leads to the conclusion that T_{00} likewise cannot depend on second x^0-derivatives of $g_{\mu\nu}$. The demonstration above that gravity is attractive depends then only on one of the four constraint equations.

[22] It needs to be stressed that the dimensions of the ball must be small in relation to the radius of curvature, so that comparison between the velocities of the balls is meaningful.

[23] The discussion so far has followed that of Baez and Bunn (2004) and Carroll (2004), pp. 167–70.

Curvature and the Weyl tensor

A word about the field equations and curvature. The Ricci tensor is the trace of the Riemann curvature tensor: $R_{\mu\nu} = R^{\alpha}{}_{\mu\alpha\nu}$. The trace-free part of the curvature tensor is therefore not algebraically related to the stress-energy tensor (at least by the Einstein field equations) and depends on boundary conditions. So the vacuum field equations, for instance, have solutions that correspond to flat Minkowski space-time and ones in which gravitational waves are propagating. These solutions differ only in relation to the trace-free part of the Riemann tensor that is captured by the so-called Weyl tensor which in four-dimensional space-time takes the form

$$C_{\rho\sigma\mu\nu} = R_{\rho\sigma\mu\nu} + \frac{1}{3}g_{\rho[\mu}g_{\nu]\sigma}R - g_{\rho[\mu}R_{\nu]\sigma} + g_{\sigma[\mu}R_{\nu]\rho}, \tag{9.18}$$

where antisymmetrization in a pair of indices is indicated by surrounding them by square brackets, e.g. $A_{[\mu}B_{\nu]} = A_{\mu}B_{\nu} - A_{\nu}B_{\mu}$. The Weyl tensor may not be algebraically related to $T_{\mu\nu}$, but because of the Bianchi identities and the field equations, a first-order differential equation relates these tensors, which is analogous to Maxwell's equations in electrodynamics.[24]

Matter and space

The standard terminology in which we distinguish between $g_{\mu\nu}$ and 'matter' fields is arguably misleading, as is the oft-repeated claim that $g_{\mu\nu}$ plays a dual role in GR, acting both as the gravitational potential and the form of the background space-time, or arena, within which physical processes play themselves out. The fact that $g_{\mu\nu}$ cannot vanish anywhere in spacetime makes it unlike any other physical field, as does the fact that it couples with every other field. Gravity is different from the other interactions, but this doesn't mean that it is *categorically* distinct from, say, the electromagnetic field. Carlo Rovelli put the point this way in 1997:

> A strong burst of gravitational waves could come from the sky and knock down the rock of Gibraltar, precisely as a strong burst of electromagnetic radiation could. Why is the ... [second] 'matter' and the ... [first] 'space'? Why should we regard the ... [first] burst as ontologically different from the second? Clearly the distinction can now be seen as ill-founded.[25]

Conflating $g_{\mu\nu}$ with space-time itself, or an essential part of it, has also given rise to the widespread misapprehension that the existence of physical solutions of the vacuum field equations, including the Minkowski metric $\eta_{\mu\nu}$ with its preferred global inertial coordinate systems, is inconsistent with Mach's principle. Once it is recognized that $g_{\mu\nu}$ is an autonomous field (or fields) in its own right, nothing in

[24] For further details, see Carroll (2004), pp. 169–70.

[25] Rovelli (1997), p. 193.

Mach's philosophy implies that its dynamical behaviour, or very existence, must be determined by the presence of (other kinds of) matter.

Rovelli contrasts two ways of interpreting Einstein's 'identification' of the gravitational field and geometry: as (a) the discovery that the gravitational field is nothing but a local distortion of space-time geometry or as (b) 'the discovery that *space-time geometry is a manifestation of a particular physical field*, the gravitational field'. Rovelli prefers (b), reserving the term 'space-time' for the unobservable differential manifold.

> Physical reality is now described as a complex interacting ensemble of entities (fields), the location of which is only meaningful with respect to one another. The relation among dynamical entities of being *contiguous* ... is the foundation of the spacetime structure. Among these various entities, there is one, the gravitational field, which interacts with every other one and thus determines the relative motion of the individual components of every object we want to use as rod or clock. Because of that, it admits a metrical interpretation.[26]

The first two sentences of this passage express a view very similar to one Einstein defended, and which was reinforced in section 9.2.2 above. The remaining sentences, suggestive but perhaps a little vague, lead nicely on to our next point.

The chronometric meaning of $g_{\mu\nu}$

Nothing in the form of the equations *per se* indicates that $g_{\mu\nu}$ is the metric of space-time, rather than a (0, 2) symmetric tensor which is assumed to be non-singular, but, significantly, whose signature is indeterminate. As James Anderson wrote in 1967

> [F]rom the point of view of the principle of general invariance we need not interpret $g_{\mu\nu}$ as a metric, nor $R_{\mu\nu}$ as a Ricci tensor. ... [Einstein's field equations] do not rest on such an interpretation; one can show that they are the only dynamical equations of second differential order for a symmetric tensor $g_{\mu\nu}$ that are in accord with the principle of general invariance as we have interpreted it. ... As in all physical theories we will look for consequences of ... [the field equations] that will lead us to associate $g_{\mu\nu}$ with some observable element of the physical world.[27]

The 'chronogeometric', or 'chronometric', significance of $g_{\mu\nu}$ is not given a priori. How does it come about that $g_{\mu\nu}$ is surveyed by physical rods and clocks, and that its null and time-like geodesics are associated with the world-lines of photons and massive particles respectively? It will be argued below that it is only the geodesic motion of massive particles that can be read more or less directly off from the general form of the field equations. The rest of the operational significance of $g_{\mu\nu}$ depends on other considerations.

[26] op. cit., p. 194.   [27] Anderson (1967), p. 342.

9.3 TEST PARTICLES AND THE GEODESIC PRINCIPLE

The first empirical evidence that Einstein had of the validity of his field equations had to do with the anomalous advance of the perihelion of the planet Mercury. The extraordinary emotion Einstein felt at the moment in 1915 when he accounted for the tiny but nagging discrepancy between Mercury's orbit and the Newtonian predictions is well documented. Using what was an approximation to the still-to-be-discovered exact Schwarzschild solution of the vacuum field equations, Einstein treated Mercury as a test particle and assumed that its motion corresponds to a geodesic of the metric.

From his earliest inklings of a theory of gravity based on the principle of equivalence and the curvature of space-time, until 1927, Einstein assumed that all test bodies would follow the grooves or ruts of space-time defined by curves that are straight, or equivalently that are of extremal length. We have seen that during this period Einstein assigned a causal role to space-time structure in precisely this sense: to nudge the particles along such privileged ruts. This kind of action of space-time on matter was taken to be primitive; fortunately it turned out to be unnecessary. Appeal to the form of the field equations was enough to deliver the principle of geodesic motion.

In one of the most eloquent sections of their monumental book on gravitation, C. Misner, K. Thorne, and J. Wheeler give a detailed, if informal account[28] of the way in which the problems of defining and treating test bodies, and defining the background geometry, can be handled in GR. In particular, they explain how the geodesic behaviour of test bodies can be derived from the vanishing of the covariant divergence of the stress-energy tensor associated with the body:

$$T^{\mu}{}_{\nu;\mu} = 0. \tag{9.19}$$

Recall that these equations are an immediate consequence of the field equations (9.12), since the covariant divergence of the Einstein tensor vanishes identically (see (9.7) above)[29]. (Although it is common for (9.19) to be referred to as a 'conservation law', it was appreciated from the very beginning that the presence of the covariant derivative, and not the simple partial derivative in the equation makes this reading strictly untenable in curved space-time.)

Some version of the derivation of the geodesic equation of motion for test bodies from (9.19) is now a common feature of textbooks on GR, as is recognition of its limitations: that corrections to this equation arise from the interaction of the spin of an object of finite dimensions with the background space-time curvature.

[28] See Misner *et al.* (1973), section 20.6, pp. 471–80. References to more systematic treatments are given on p. 480.

[29] The connection between (9.12) and the requirement of general covariance will be spelt out in the Appendix A.

The first realization that such a derivation of geodesic motion is possible is sometimes attributed[30] to Einstein and Grommer in 1927. (Einstein's approach in this and later work was to treat elementary particles as field singularities. Different approaches to the derivation of the geodesic theorem were adopted by other workers.)[31] But the first derivation of geodesic motion is found as early as 1918 in a review paper on GR by Eddington.[32] Five years later, Eddington stressed that it is a 'blemish' in a 'deductive theory' simply to assume that test bodies—particles and light rays—trace out geodesics and null geodesics respectively of the metric, and he provided an interesting plausibility argument based on the equations (9.19) for the geodesic motion of 'symmetrical' particles.[33]

The fact that geodesic motion is a theorem and not a postulate has striking consequences that cannot be overemphasized. Earlier in the book I argued that it (and the need for corrections in the case of bodies with spin) casts doubt on the widespread view that space-time structure, in and of itself, can act directly on test bodies. But it also sheds light on the meaning of inertia (in the sense of inertial motion) and gravity.

Gravity traditionally has had two faces. It explains why things fall down and it explains why the fall is not uniform across space and time. Take the latter aspect first. Two objects are in free-fall in my office: the head of my student who has just nodded off in a tutorial, and the copy of Misner, Thorne, and Wheeler's *Gravitation* that I am throwing at him. Ignoring initial horizontal components of the motions, the two objects do not quite move along parallel lines, they are each heading after all towards the centre of the Earth. This of course has to do in GR with geodesic deviation, with the so-called 'tidal' effects of space-time curvature. It is tempting to think that in GR this is all we mean, or should mean, by gravity. For when we come to explaining why the objects are falling in the first place, the answer seems almost banal. It is because the frame defined by my office is accelerating in relation to the local inertial frames, and in relation to the latter frames the objects, being force-free, are simply moving inertially.[34] The glory of

[30] See Adler *et al.* (1975), p. 352, and Misner *et al.* (1973) p. 480.

[31] See in particular Fock (1969).

[32] A fascinating detailed account of early, independent work on the equation of motion in GR by Eddington, de Donder, Weyl, Pauli, and several others, which not only predated the collaboration between Einstein and Grommer but was never properly acknowledged by Einstein, is found in Havas (1989).

[33] In his 1923 book *The Mathematical Theory of Relativity*, Eddington treated a particle as a narrow tube containing non-zero $T_{\mu\nu}$, with momentum and mass obtained by integrating $T_{4\nu}$ over a three-dimensional volume. He realized that the constancy along of the tube of the 'dynamical' velocity 4-vector does not show that the direction of this vector is in the direction of this tube. But he argued that if the particle is symmetrical (or at least defines three perpendicular planes of symmetry) there would be no preferential direction in which the momentum can point to cause deviation from a geodesic. See Eddington (1965), pp. 125, 127. It should be noted that in the same publication (section 74c), Eddington also gave a careful treatment of the behaviour of light wave fronts based on the generally covariant form of Maxwell's equations.

[34] This argument doesn't of course account for why gravitation is attractive, an issue we dealt with earlier.

this explanation is that it accounts for the universality of free fall, that all bodies, independent of their constitution, fall at the same rate. But we must be especially careful when we conclude that gravity is either merely a pseudo-force, or not a force at all.

It is true that no force in the traditional sense is acting on a freely falling body, because such a force strictly only has meaning in GR in relation to the non-gravitational interactions, whose job it is to pull objects with suitable attributes like charge etc. off the geodesics. It is the notion found in SR, not quite the same as a Newtonian force, but clearly related to it. But when 'force-free' test bodies undergo geodesic flow in GR, whether there is geodesic deviation or not, such motion is ultimately due to the way the Einstein field $g_{\mu\nu}$ couples to matter, as determined by the field equations. It is a consequence, even if one that remained hidden for some years after the birth of GR, of the fundamental dynamics of the theory. As Misner, Thorne, and Wheeler stress:

> Only here [in GR] does the conservation of energy-momentum appear as a fully automatic consequence of the of the inner working of the machinery of the world ... It makes no difficulties whatsoever for Maxwell's equations that one had shifted attention from a world line that follows the Lorentz equation of motion to one that does not. Quite the contrary is true in general relativity. To shift from right world line (geodesic) to wrong world line makes the difference between satisfying Einstein's field equations in the vicinity of that world line and being unable to satisfy Einstein's field equations.[35]

Inertia, in GR, is just as much a consequence of the field equations as gravitational waves. For the first time since Aristotle introduced the fundamental distinction between natural and forced motions, inertial motion is part of the dynamics. It is no longer a miracle.[36]

9.4 LIGHT AND THE NULL CONES

In the majority of experimental tests of GR, light, or electromagnetic radiation more generally, plays a crucial role as a tracer, mapping out the curvature of space-time. These tests involve the gravitational 'bending' of light, the gravitational redshift (though this is a test not of the existence of curvature but of the equivalence principle), the time delay of radar pulses in the gravitational field of the Sun, and most recently the gravitational lensing (and microlensing of starlight) in the gravitational field of galaxies.

In all these experiments, light is treated as a test body, materializing null geodesics in prescribed gravitational fields that are solutions of the Einstein vacuum equations (9.6)—the Schwarzschild metric in the case of the solar system.

[35] Misner *et al.*, p. 475. For a more critical evaluation of the geodesic theorem in GR, see Tavakol and Zalaletdinov (1998).

[36] Another way of putting the point is that there are no truly force-free bodies in GR; see Trautman (1966) and Anderson (1967), p. 438.

(The possible novel effects of the coupling between the Einstein and Maxwell fields such as the generation of electromagnetic waves by gravitational waves are too small to be detected.) But, just as with test bodies and their geodesic motion, it would be wrong to imagine that the behaviour of light is defined in GR by the null cones. What is needed is a determination of the behaviour of light in a curved space-time as given by the best theory of the Einstein–Maxwell equations, by way of the geometric optics approximation. This has long been appreciated (possibly first by Eddington)[37] and the argument usually goes like this.

The standard Lagrangian density for the Maxwell field in the absence of matter sources is given by

$$\mathcal{L}_{\mathrm{EM}} = -\frac{1}{4}F_{\mu\nu}F^{\mu\nu}\sqrt{-g}. \tag{9.20}$$

the field strength tensor being given by $F_{\mu\nu} \equiv A_{\nu;\mu} - A_{\mu;\nu} = A_{\nu,\mu} - A_{\mu,\nu}$ where A_μ is the electromagnetic four-potential. The Maxwell field couples to gravity because in (9.20) $F^{\mu\nu} \equiv g^{\mu\sigma}g^{\nu\lambda}F_{\sigma\lambda}$. The Lagrangian (9.20) fits in to the RHS of (9.11), and the metric will be a solution of the full field equations (9.12). However, we may assume that in the experiments we are interested in, the effect of the light beam on the background metric is negligible, and we treat the metric as a solution of the vacuum equations (9.6). If we apply Hamilton's principle with respect to arbitrary variations in the A_μ we obtain in the standard way the generally covariant field equations

$$F^{\mu\nu}{}_{;\mu} = 0. \tag{9.21}$$

The further familiar equations

$$F_{\mu\nu;\sigma} + F_{\nu\sigma;\mu} + F_{\sigma\mu;\nu} = 0 \tag{9.22}$$

follow of course from the definition above of $F_{\mu\nu}$ in terms of the A_μ.

In the geometric optics regime,[38] wave solutions to the equations (9.11) are considered where the wavelength is much smaller than the scale on which either the amplitude or the background space-time curvature is varying. Locally, one can think of the wave as an undistorted plane wave. So we suppose that the 4-potential can be written as

$$A_\mu = \mathrm{Re}\left\{\left[a_\mu(x) + O(\epsilon) + O(\epsilon^2) + \cdots\right] \exp\left(\frac{i}{\epsilon}\theta(x)\right)\right\}, \tag{9.23}$$

where the amplitude a_μ is complex and slowly varying, and the phase θ is real and rapidly varying. The dimensionless parameter ϵ tends to zero as the typical

[37] See fn. 33 above.

[38] A useful treatment is found in Misner *et al.* (1973), section 22.5; see also Ehlers (1973).

wavelength of the electromagnetic signal becomes shorter and shorter; the electromagnetic field equations can be solved order-by-order in ϵ. The wave vector k_μ is defined as the gradient of the phase

$$k_\mu \equiv \theta_{,\mu}, \tag{9.24}$$

and light rays are defined to be the curves normal to the surfaces of constant phase θ. Thus the differential equation for a light ray is

$$\frac{dx^\mu}{ds} = k^\mu = g^{\mu\nu}\theta_{,\nu}. \tag{9.25}$$

Now insert the potential (9.23) into the source-free field equations (9.21) and collecting terms of order $(1/\epsilon^2)$ one derives first that

$$k^2 \equiv k^\mu{}_\mu = 0, \tag{9.26}$$

which establishes that k_μ is a null vector. If one now takes the gradient of k^2, and recalling that the wave vector is the gradient of a scalar, it can be shown that

$$k^\mu k_{\mu;\nu} = k^\mu k_{\nu;\mu} = 0, \tag{9.27}$$

which means that the curve for which k^μ is its tangent parallel transports k^μ along itself: the curve is a geodesic. But from (9.25) the curve is just the path of a light ray, so it has been shown that light rays propagate along null geodesics in the geometric optics approximation. Light acts as a tracer of the conformal structure associated with the metric field, not because it must but because the electromagnetic field equations say it does.

9.4.1 Non-minimal Coupling

But interest has been growing in recent years in possible modifications to the standard electromagnetic Lagrangian density (9.20). Consider for example the Lagrangian $\tilde{\mathcal{L}}_{\text{EM}} =$

$$\left(-\frac{1}{4}F_{\mu\nu}F^{\mu\nu} + \frac{1}{4}\xi R F_{\mu\nu}F^{\mu\nu} + \frac{1}{2}\eta R_{\mu\nu}F^{\mu\rho}F^{\nu}{}_{\rho} + \frac{1}{4}\zeta R_{\mu\nu\rho\sigma}F^{\mu\nu}F^{\rho\sigma}\right)\sqrt{-g}, \tag{9.28}$$

where $F_{\mu\nu}$ is defined as above, $R_{\mu\nu\rho\sigma}$ is the curvature tensor and ξ, η, and ζ are constants having dimensions [length]^2. This Lagrangian density brings together a number of terms that individually have appeared in various investigations of the Einstein–Maxwell equations that follow from non-minimal coupling (the meaning of which is discussed in the following section), but which reduce to the standard equations when space-time is Minkowskian and hence flat. The case of

$$\eta = -2\xi, \quad \zeta = \xi \tag{9.29}$$

is of particular interest. A 1976 theorem due to G. W. Horndeski[39] purports to show that the Lagrangian density (9.18) with the constants given by (9.29) give rise to the most general electromagnetic equations which are (i) derivable from a variational principle, (ii) at most second-order in the derivatives of both $g^{\mu\nu}$ and A_μ, (iii) consistent with charge conservation, and (iv) compatible with Maxwell's equations in Minkowski space-time.

Satisfaction of the conditions (i) to (iv) above is sometimes taken to establish compatibility with the currently accepted principles of electromagnetism.[40] But a word is in order at this point regarding condition (iv). The only justification for this condition that I can see is the notion that in the special relativistic limit, i.e. in the absence of curvature, the traditional Maxwell's equations must be recovered. This rationale is questionable. It rests on the 'counterfactual' notion of limit that was raised and criticized in section 6.5 above. The special relativistic limit of GR is a subtle business, as we see in the next section, but what it involves is not postulating a globally flat Minkowski space-time (which is consistent with the Einstein field equations only if the stress energy tensor $T_{\mu\nu}$ vanishes everywhere, so there is strictly no matter) but rather the specification of experimental conditions under which the observed effects of space-time curvature are negligable. The interesting examples of non-minimal coupling are cases where this condition fails.

In 1980, I. T. Drummond and S. Hathrell calculated the effective action in QED with a background gravitational field, or more specifically the one-loop vacuum polarization contribution to the action. Vacuum polarization in QED confers a size to the photon of the order of the Compton wavelength of the electron, defined as $\lambda_C = \hbar/m_e c$. As in the case of the spinning particle of finite dimensions that violates the principle of geodesic motion, the motion of the photon can now be influenced by a tidal gravitational effect that depends on the curvature. In the one-loop approximation, Drummond and Hathrell[41] obtained a Lagrangian density of the form (9.28) but with an extra additive term

$$\mathcal{L}_{\text{QED}} = \tilde{\mathcal{L}}_{\text{EM}} + \lambda F^{\mu\sigma}{}_{;\mu} F^{\nu}{}_{\sigma;\nu}, \tag{9.30}$$

where the coefficients in (9.28) are (compare with (9.29) for the Horndeski theorem) in units $\hbar = c = 1$

$$\xi = -\frac{\alpha}{144\pi}\lambda_C^2, \quad \eta = \frac{13\alpha}{360\pi}\lambda_C^2, \quad \zeta = -\frac{\alpha}{360\pi}\lambda_C^2, \tag{9.31}$$

[39] See Horndeski (1976), Horndeski and Wainwright (1977), Buchdahl (1979), and Müller-Hoissen and Sippel (1988).

[40] See Teyssandier (2003).

[41] Drummond and Hathrell (1980).

α being the fine-structure constant. Note that the additive term in (9.30) does not depend on curvature; it represents off-mass-shell effects in vacuum polarization but is irrelevant for considerations related to the speed of photons.

Indeed, in the low frequency approximation $\lambda \gg \lambda_C$, it can be shown that the analogue of (9.26) is

$$k^2 - \frac{13\alpha\lambda_C^2}{180\pi} R_{\mu\nu}k^\mu k^\nu - \frac{\alpha\lambda_C^2}{45\pi} R_{\mu\nu\lambda\rho}k^\mu k^\lambda a^\nu a^\rho = 0, \qquad (9.32)$$

where a_μ is the amplitude defined in (9.23), the normalized part of which is the polarization of the wave. It is helpful to rewrite this equation using the field equations (9.12), so that the matter and gravitational contributions are separated:[42]

$$k^2 - \frac{22\alpha\lambda_C^2}{45} T_{\mu\nu} - \frac{\alpha\lambda_C^2}{45\pi} C_{\mu\nu\lambda\rho}k^\mu k^\lambda a^\nu a^\rho = 0, \qquad (9.33)$$

where $C_{\mu\nu\lambda\rho}$ is the Weyl tensor, and the gravitational constant G is set equal to 1. The second term in (9.33) takes into account the effect of any one of a number of distinct backgrounds on photon propagation. Examples that have been studied in detail include (classical) electromagnetic fields, Casimir cavities involving parallel conducting mirrors, and finite temperature environments. In short, QED appears to predict, without any appeal to gravity, deviation from null-cone behaviour for photons under certain conditions.[43] The deviations are tiny, far beyond the reach of current experimental techniques.

We are more familiar with the fact that the velocity of light may differ from c inside a transparent material medium, which may or may not be dispersive. What is striking in the case of the electromagnetic field environment is that light is now scattering off light: the effective action (traditionally known as the Euler–Heisenberg action) introduces non-linearity into the Maxwell field equations. In the case of the Casimir cavity, it is the reflective boundaries that provide a kind of medium in otherwise empty space, one which alters the velocity of light. But what is truly striking in this case is the prediction that photons propagating perpendicular to the conducting plates do so 'superluminally': the *Scharnhorst effect.*

The third term in (9.33) is of course solely gravitational, and because it depends on a_μ, the interaction produces a polarization-dependent shift in the velocity of light, or what is called *gravitational birefringence.* Now because of the symmetry properties of the Weyl tensor, it turns out that the two physical polarizations contribute equally in magnitude, but oppositely in sign. The upshot is that for

[42] See Shore (2003), §4.1.

[43] See, e.g., Latorre *et al.* (1995), in which a unification of results concerning variations in the speed of low-energy photons due to modifications of the vacuum is given.

space-times with vanishing Ricci tensor (which means that the only interaction term is the one containing the Weyl tensor) if one polarization produces a time-like deviation from the null-cone behaviour, the other must produce a space-like deviation. Thus, superluminal velocities are predicted in this case too.

Are such theories self-consistent? It appears so: detailed studies of the causality problem have shown that no clear instances of the so-called grandfather paradox arise in these theories.[44] But despite their intriguing nature, these theories are based on an approximation which is known not to be the whole story. So far the effective actions hold only in the low-frequency regime, while it is known that for the purposes of considerations related to causality the relevant concept (particularly in dispersive media) is wavefront velocity, which in turn can be shown to be the high frequency limit of the phase velocity.[45] To deal with this case, a generalization of the Drummond–Hathrell effective action containing higher orders of derivatives of the fields is required, which indeed leads to dispersive propagation of light generally. A careful treatment involving the generalized action has been undertaken recently by Graham Shore in the special case of Bondi–Sachs space-time, and it was found that here exceptionally the dispersion vanishes so that the wavefront velocity coincides with the superluminal low-frequency phase velocity. 'This is potentially a very important result. It appears to show that there is at least one example in which the wavefront truly propagates with superluminal velocity. If so, quantum effects would indeed have shifted the light cone into the geometrically spacelike region.'[46]

But alas, the situation is not entirely clear-cut, as Shore himself stresses. The problem is that in the discussion of light propagation in background magnetic fields based on the Euler–Heisenberg action, it is known that the perturbative approach hitherto discussed is not adequate to deal with the full range of frequencies in the effectively dispersive medium. And it is simply not clear whether the gravitational case should be any different, and so whether ignoring a non-perturbative contribution to the equation for light propagation is justified. If it isn't, it seems that the wavefront velocity may after all be driven back to c, as it is in the case of background magnetic fields in the non-perturbative analysis.[47]

The technical details involved are far too intricate to go into here. My motivation in mentioning this recent work is merely to point out that no rigorous derivation that the wavefront velocity *is* strictly c in the exotic gravitational regime under discussion is available, even if it might be supposed that this identification does, in the end, hold. What Shore and others have demonstrated is that quantum effects *may* lift light off the null cones, and that even if superluminal speeds were reached (as is predicted in the Casimir cavity) no fundamental conceptual inconsistencies are necessarily at play. What would be violated is not logic, but the strong equivalence principle.

[44] See, e.g., Liberati *et al.* (2002) and Shore (2003).
[45] See Shore (2003), §6.2.
[46] op. cit., §6.4.
[47] ibid.

9.5 THE STRONG EQUIVALENCE PRINCIPLE

9.5.1 The Local Validity of Special Relativity

The apparent fact that the trajectory of a freely falling body does not depend on its internal structure and constitution was of singular importance for Einstein in his development of GR, as every textbook on the subject attests. Indeed, in modern formulations of the theory this fact is standardly elevated to the status of a principle: the *weak equivalence principle* (WEP), which we discussed in Chapter 2. (The terminology is of course a consequence of the fact that in the context of Newtonian dynamics, the principle demands that the inertial mass of the body is proportional to its weight, or passive gravitational mass.) The principle follows from the above-mentioned derivation of the geodesic principle from the field equations to the extent that the particular constitution of the test body is irrelevant to the derivation.

A more far-reaching claim is the *strong equivalence principle* (SEP), which will be defined here as follows.[48] There exist in the neighbourhood of each event preferred coordinates, called locally inertial at that event. For each fundamental non-gravitational interaction, to the extent that tidal gravitational effects can be ignored the laws governing the interaction find their simplest form in these coordinates. This is their *special relativistic form*, independent of space-time location.[49] Note that in practice, any measurements that are being performed involving these interactions must be such that their action on the sources of the gravitational field can be ignored.

The SEP, as just defined, does not entirely specify what the local inertial coordinates are. In standard GR, they are connected to the metric in the following way. There exist at each point (event) p locally geodesic coordinates. These are coordinates such that at p, the components of the symmetric connection compatible with $g_{\mu\nu}$ vanish: $\Gamma^{\alpha}{}_{\beta\nu} = 0$. A subset of such coordinates can be found such that all geodesics through p look like straight lines, and at p the first (ordinary) derivatives of $g_{\mu\nu}$ vanish: $g_{\mu\nu} = \eta_{\mu\nu}$ and $g_{\mu\nu,\lambda} = 0$. For points in the neighbourhood of p, the difference in the value of the components of the metric from that at p depends on the Riemann curvature tensor (which of course depends on second derivatives of $g_{\mu\nu}$ and hence need not vanish at p). Thus, as long as the curvature tensor, or 'tidal field', evaluated in the mentioned orthonormal frame at p, is 'sufficiently small', the coordinates of points in this neighbourhood will differ little from the inertial coordinates for events on the tangent space at p related to

[48] The discussion will closely follow that in Ehlers (1973).

[49] Sometimes what we are calling the SEP is referred to as the Einstein equivalence principle (EEP), and the term 'strong equivalence principle' is reserved for the principle that deals with bodies with gravitational self-energy. (See Will (2001), §3.1.2, and Carroll (2004), p. 50.) For a criticism of this distinction, see Ohanian and Ruffini (1994), p. 56. A recent treatment of the strong equivalence principle in the philosophical literature is in Ghins and Budden (2001).

the points in the manifold by the so-called exponential map.[50] Since the tangent space is Minkowskian, it might be thought that the special relativistic chronometric significance of of the Minkowski metric in SR is automatically recovered locally in GR. But it is only through the SEP that such chronometric significance can be given to the tangent space geometry in the first place.[51] This still leaves the irksome question: what does 'sufficiently small' mean? Is the curvature of space-time in the vicinity of the Earth small enough to make a freely falling laboratory the size of the room you are presently in realize a local inertial frame? Whether you can detect tidal effects in a space the size of the room depends on what kind of equipment you have access to, or in some cases how much time you have at your disposal![52] And the issue of time reminds us that the closest thing we have in GR to inertial coordinate systems in classical mechanics and SR are not the local inertial systems as defined above, but non-rotating *freely falling frames* (otherwise known as Fermi normal coordinates, or proper reference frames of a freely falling observer). These are local inertial frames maintained along time-like geodesics, whose coordinate axes are parallel transported (in the sense of the metric compatible connection) along the geodesic.[53]

James L. Anderson correctly emphasized that the SEP as we have defined it contains two distinct principles.[54] The first, let us call it SEP_1, is that measurements on *any* physical system will serve (approximately) to determine the *same* affine connection in a given region of space-time (which Anderson called the equivalence principle). In other words, the non-gravitational interactions all pick out the same local inertial structure. The second principle, SEP_2, is that only the connection determined by $g_{\mu\nu}$, with its Lorentzian signature, appear in the dynamical laws for these systems, or that in a 'sufficiently small' region of space-time, the laws of special relativity are valid. Anderson claimed that the first principle is vital to GR, but not the second.

In relation to SEP_2, what does it mean that in freely falling frames 'the non-gravitational laws of physics are those written in the language of special relativity'?[55] What is this language? If we must be clear about the content of SR, or at least that part of it concerned with space-time structure, it is surely here. What is being assumed is that the covariance group of the equations is the inhomogeneous Lorentz group, which the *minimal coupling* condition ensures. This is the claim that the matter fields do not couple to the Riemann curvature tensor or its contractions. Recall that in SR, inertial frames are global, which implies that the curvature vanishes everywhere, and hence trivially makes no appearance in the

[50] For more details see Ehlers (1973).

[51] The present section is an attempt to spell out this claim, which was made briefly in Brown and Pooley (2001) in response to the earlier claim by Torretti (1983) that the local validity of SR is automatic in the above sense.

[52] See Ohanian and Ruffini (1994) §1.9, and Hartle (2003), example 6.3, p. 120.

[53] For further details, see e.g. Ehlers (1973) §2.10, Hartle (2003), §20.5, and Cook (2004).

[54] Anderson (1967), §10–2. [55] Will (2001), §2.1.

laws of physical interactions. This feature is now absorbed into GR in the requisite local context. It is worth emphasizing that minimal coupling is sufficient for local Lorentz covariance, but is not necessary, at least in the sense of the homogeneous group. Consider for example the propagation of a free scalar field with mass m in Minkowski space-time described by

$$\left(\eta^{\mu\nu}\partial_\mu\partial_\nu - m^2\right)\Phi = 0. \tag{9.34}$$

One possible generalization of (9.34) to curved space-time is

$$\left(g^{\mu\nu}\nabla_\mu\nabla_\nu - m^2 - \xi R\right)\Phi = 0 \tag{9.35}$$

where as above R is the curvature scalar and ξ is a numerical constant. Minimal coupling implies $\xi = 0$, but local Lorentz covariance doesn't. Analogously, the cases of non-minimal coupling discussed in the previous section do not involve violation of local Lorentz covariance.

Violation of local Lorentz covariance (homogeneous group) requires the introduction of further fields—typically the preferred frame is defined by a vector field or the gradient of a scalar field—over and above $g_{\mu\nu}$ and the standard matter fields. A recent example is a theory due to Ted Jacobson and David Mattingly,[56] in which the extra, time-like unit vector field u^α is itself a dynamical object. Without going into details, the total action can be written in the form $S = S_{\text{grav}} + S_{\text{vector}} + S_{\text{matter}}$, where S_{grav} is the standard Hilbert–Einstein action, S_{vector} depends on $g_{\mu\nu}$, u^α, and $\lambda = \lambda(x)$ (a Lagrange multiplier associated with the normalization of u^α), and S_{matter} depends on $g_{\mu\nu}$, u^α, and ψ, where ψ is the generic matter field. I take it that such a theory is also incompatible with minimal coupling as it is standardly construed.

Minimal coupling is usually identified with the so-called 'comma-goes-to semicolon' rule. One starts with a Lorentz covariant theory of some non-gravitational force and formally replaces in the equations the flat Minkowski metric $\eta_{\mu\nu}$ by the curved metric $g_{\mu\nu}$, and the ordinary derivative represented by a comma with the covariant derivative represented by a semicolon. The resulting theory is automatically consistent with the local validity of SR. Of course textbooks almost invariably warn that such a procedure for generating theories consistent with minimal coupling is only unambiguous in theories in which the field equations are first order. In practice, factor ordering problems are easily avoided, and anyway the rule is essentially heuristic.[57] A more serious limitation of minimal coupling itself

[56] See Jacobson and Mattingly (2001); this paper contains references to several earlier gravitational theories incorporating violation of local Lorentz covariance by other authors.

[57] The ambiguity in the comma-goes-to-semicolon rule for second-order equations is reminiscent of the ambiguity in the process of canonical quantization in mechanics, which arises when dynamical systems can be associated with 'inequivalent' Lagrangians, i.e. Lagrangians that are not related by a divergence. See, e.g., Morandi *et al.* (1990).

is illustrated by the example of the scalar field associated with equation (9.35). Normally, it is argued that conformal invariance results in the factor ξ taking the value $1/6$, whereas as mentioned above minimal coupling requires $\xi = 0$. But if it is required merely that the physical properties of wave propagation rather than the form of the equation reduce locally to those of SR, then the value $1/6$ is permissible.[58] And it is known that the equation of motion of the spin of a rigid body fails to satisfy minimal coupling, because tidal torques are present in the correct equation and the motion of the spin depends on the Riemann curvature tensor.[59]

The correctness of Anderson's claim that minimal coupling, and hence the local validity of SR, is not essential to the programme of GR naturally depends on what that programme is taken to be. At the very least, it is clear that minimal coupling is not a direct consequence of the form of Einstein's field equations.[60]

9.5.2 A Recent Development

It is widely known that there is a route to GR based on global Minkowski space-time with spin-2 gravitons. It involves postulating a symmetric tensor field $h_{\mu\nu}$ propagating in flat space-time according to a certain action principle. (The action happens to be that associated with the so-called linearized Einstein tensor.) If it is then required that $h_{\mu\nu}$ couple to its own stress-energy tensor, as well as to the matter stress-energy tensor, an iterative process involving induced higher-order terms in the action leads to the Hilbert–Einstein action (or (1.3) with $\lambda = 0$) for $g_{\mu\nu} \equiv \eta_{\mu\nu} + h_{\mu\nu}$, with matter also in the end coupling to $g_{\mu\nu}$.[61] The flat background metric in this approach becomes bereft of any direct operational meaning because the SEP holds for the local inertial frames defined relative to $g_{\mu\nu}$ and not to $\eta_{\mu\nu}$.

An interesting, and potentially more significant, variation of this theme concerns a recent proposal to modify GR. In 2004, Jacob D. Bekenstein introduced a new covariant theory of gravity,[62] the latest in a line of theories that owe their origin and motivation to the work of Mordehai Milgrom in the 1980s. Milgrom suggested a modification of Newtonian gravitational dynamics (MOND), based on a new acceleration scale, that would account for galactic rotation curves without appealing to dark matter.[63] Bekenstein's new theory, denominated T*e*V*e*S (for

[58] Sonego and Faraoni (1993).

[59] See Ohanian and Ruffini (1994), pp. 379–80. A very interesting discussion of the apparent violation of the strong equivalence principle by particle detectors in quantum field theory on curved background space-times is found in Sonego and Westman (2003). A discussion of the strong equivalence principle from the perspective of quantum processes is also found in Audretsch *et al.* (1992).

[60] A useful discussion of non-minimal coupling of matter and the gravitational field is found in Goenner (1984).

[61] See, e.g., Carroll (2003), p. 299.

[62] Bekenstein (2004*b*), Bekenstein (2004*a*).

[63] A useful introduction to MOND is found in Milgrom (2001). A good treatment of the strengths and weaknesses of successive theories in the MOND paradigm is given in Bekenstein (2004*b*), in

'Tensor-Vector-Scalar'), postulates as its fundamental dynamical entities a metric field $g_{\mu\nu}$, a time-like four-vector field $\mathcal{U}_\mu$ and a scalar field ϕ, as well as a non-dynamical scalar field σ.

Besides its success with the dark-matter-free treatment of observed galactic rotation behaviour, T*e*V*e*S is consistent with the results of all the solar systems tests of GR, predicts gravitational lensing in agreement with the observations, and provides a formalism for constructing cosmological models. It seems too complicated to be the last word, but it deserves our attention. In particular, the structure of the theory speaks to the concerns of this book.

It is assumed that the $g_{\mu\nu}$ field has a well-defined inverse $g^{\mu\nu}$ and that $g^{\mu\nu}\mathcal{U}_\mu\mathcal{U}_\nu \equiv \mathcal{U}^\nu\mathcal{U}_\nu = -1$. The purely gravitational part of the total action is built out of the 'metric' field $g_{\mu\nu}$ just as it is in GR, but otherwise the role and meaning of this field is quite different. A second metric field, $\tilde{g}_{\mu\nu}$, is obtained from $g_{\mu\nu}$ by stretching it in the space-time directions orthogonal to $\mathcal{U}^\mu$ by a factor $e^{-2\phi}$, and shrinking it by the same factor in the direction parallel to $\mathcal{U}^\mu$ (so that it is not conformal to $g_{\mu\nu}$):

$$\tilde{g}_{\mu\nu} = e^{-2\phi}(g_{\mu\nu} + \mathcal{U}_\mu\mathcal{U}_\nu) - e^{2\phi}\mathcal{U}_\mu\mathcal{U}_\nu \tag{9.36}$$

$$= e^{-2\phi}g_{\mu\nu} - 2\mathcal{U}_\mu\mathcal{U}_\nu \sinh(2\phi). \tag{9.37}$$

It can be shown that like $g_{\mu\nu}$, $\tilde{g}_{\mu\nu}$ has a well-defined inverse.

As mentioned, the purely gravitational action essentially coincides with that of standard GR (9.3) for $g_{\mu\nu}$:

$$S_{\text{grav}} = \frac{1}{16\pi G}\int_\Omega R\sqrt{-g}\,\mathrm{d}^4x. \tag{9.38}$$

The action for the pair of scalar fields is

$$S_{\text{scalar}} = -\frac{1}{2}\int_\Omega \left[\sigma^2 h^{\mu\nu}\phi_{,\mu}\phi_{,\nu} + \frac{1}{2}Gl^{-2}\sigma^4 F(kG\sigma^2)\right]\sqrt{-g}\,\mathrm{d}^4x, \tag{9.39}$$

where $h^{\mu\nu} \equiv g^{\mu\nu} - \mathcal{U}^\mu\mathcal{U}^\nu$ and F is a free dimensionless function. The constant positive parameters k, l are such that k is dimensionless, and l has dimensions of length. The action of the vector field $\mathcal{U}_\mu$ takes the form

$$S_{\text{vector}} = -\frac{K}{32\pi G}\int_\Omega \Big[g^{\alpha\beta}g^{\mu\nu}\mathcal{U}_{[\alpha,\mu]}\mathcal{U}_{[\beta,\nu]} - 2(\lambda/K)(g^{\mu\nu}\mathcal{U}_\mu\mathcal{U}_\nu + 1)\Big]\sqrt{-g}\,\mathrm{d}^4x, \tag{9.40}$$

which it is stressed that much of the relevant galactic observational data post-dates the original predictions of MOND.

where $\lambda = \lambda(x)$ is a Lagrange multiplier associated with the normalization of $\mathcal{U}_\mu$, and K is a dimensionless constant. Finally, the matter action associated with fields written generically as f^α is

$$S_{\text{matter}} = \int_\Omega \mathcal{L}_{\text{matter}}(\tilde{g}_{\mu\nu}, f^\alpha, f^\alpha{}_{|\mu}, \cdots)\, \mathrm{d}^4x, \tag{9.41}$$

where the covariant derivatives denoted by '$|$' are defined relative to the connection compatible with $\tilde{g}_{\mu\nu}$. Note that the Lagrangian density $\mathcal{L}_{\text{matter}}$ is to be understood as a multiple of $\sqrt{-\tilde{g}}$, which can be shown to be equal to $e^{2\phi}\sqrt{-g}$.

The T*e*V*e*S theory thus has two dimensionless parameters, k and K, as well as the constants G and l. The basic equations of the theory are obtained from the total action $S = S_{\text{grav}} + S_{\text{scalar}} + S_{\text{vector}} + S_{\text{matter}}$ by varying with respect to the fields $g^{\mu\nu}$, ϕ, σ, and $\mathcal{U}_\mu$.

We need not concern ourselves with the detailed form of these equations. Needless to say, things are a lot more complicated than in GR. In particular, the analogue of the Einstein field equations (9.12) obtained by varying with respect to $g^{\mu\nu}$ looks like

$$G_{\mu\nu} = 8\pi G\tilde{T}_{\mu\nu} + \cdots, \tag{9.42}$$

where the ellipses denotes the sum of various terms involving the ϕ, σ, and $\mathcal{U}_\mu$ fields and their first derivatives. The stress-energy tensor $\tilde{T}_{\mu\nu}$ is defined in the usual way in terms of the variational derivative of S_{matter}, but now with respect to the (inverse) 'physical' metric $\tilde{g}^{\mu\nu}$, not $g^{\mu\nu}$ as in (9.11). It turns out that in the limit $l \to \infty$, $\tilde{g}^{\mu\nu}$ coincides with $g^{\mu\nu}$, and when furthermore $K \to 0$, the theory reduces to exact GR whatever the value of k. In several familiar contexts, such as Friedmann–Robertson–Walker (FRW) cosmologies, it can be shown that standard GR predictions are obtained also in the limit $k \to 0$, $l \propto k^{-3/2}$, and $K \propto k$, whatever form the function F takes. At any rate, it is assumed in the theory that $k \ll 1$ and $K \ll 1$, and a choice of the the function F is made so as best to match observation.

The theory is not beautiful; the menagerie of dynamical fields—the price one pays for eliminating dark matter—has collectively nothing like the simple heuristic foundation that Einstein introduced for the $g_{\mu\nu}$ field. Not that dark matter is being replaced by something just like it; the way the new scalar and vector fields (whose contribution to the stress-energy tensor is usually negligable) affect the geometrical structure associated with $\tilde{g}^{\mu\nu}$ is quite different from the way dark matter is usually understood to act on $g^{\mu\nu}$. At any rate, it is far too soon to predict what the fate of the theory will be. But our concerns have more to do with conceptual than empirical matters. And the key issue is this.

The metric field that is surveyed by rods and clocks, whose conformal structure is traced by light rays and whose geodesics correspond to the motion of free bodies is clearly not $g_{\mu\nu}$, but the less 'basic' $\tilde{g}_{\mu\nu}$.[64] Indeed, it is explicitly assumed in the theory[65] that the matter Lagrangian is obtained from that associated with flat space-time $\tilde{g}_{\mu\nu} = g_{\mu\nu} = \eta_{\mu\nu}$ by using the $f^{\alpha}{}_{,\mu}$-goes-to-$f^{\alpha}{}_{|\mu}$ rule. In other words, the theory incorporates the SEP, but with respect to the $\tilde{g}_{\mu\nu}$ field.[66] It is this field, and not the metric which features in what Bekenstein curiously calls the 'geometric' part of the action S_{grav}, which acquires chronometric significance, and it does so because of the postulated dynamics in the theory. Right or wrong, the theory reminds us that the operational significance of a non-singular second-rank tensor field, and its geometric meaning, if any, depends on whether it is 'delineated by matter dynamics'.[67]

Recall Poincaré's remark cited in section 2.2.4, to the effect that if there were no rigid bodies there would be no (3-space) geometry. In the T*e*V*e*S theory, there is a lot of non-trivial dynamics associated with $g^{\mu\nu}$ (unlike $h^{\mu\nu}$ in the spin-2 graviton theory above), and hence $\tilde{g}^{\mu\nu}$, even when the matter fields f^{α} vanish everywhere. But it is the $\tilde{g}^{\mu\nu}$ field that becomes 'geometrical' in the usual four-dimensional sense when the usual matter fields are introduced.[68]

9.6 CONCLUSIONS

In the context of standard GR, Clifford Will has nicely expressed a common view:

> The property that all non-gravitational fields should couple in the same manner to a single gravitational field is sometimes called 'universal coupling'. Because of it, one can discuss the metric as a property of spacetime itself rather than as a field over space-time.

[64] The sense in which the $\tilde{g}_{\mu\nu}$ field is less basic than the field is not just that it is in terms of variations with respect to the latter that the analogue of Einstein's field equations are derived from the action principle. The FRW cosmology in the theory is of course defined in the usual way in terms of $g_{\mu\nu}$, and it is assumed that the T*e*V*e*S scalar and vector fields share the symmetries of this field.

[65] Bekenstein (2004*b*) §III.A. A recent study of the compatibility of T*e*V*e*S with observations of the cosmic microwave background and of galaxy distributions is found in Skordis *et al.* (2005).

[66] The vanishing of the covariant divergence of $\tilde{T}_{\mu\nu}$—with respect to $\tilde{g}_{\mu\nu}$—is required if the geodesic principle of free motion and the local validity of Maxwell's equations are to hold, and this requirement is not a consequence of the field equations (9.42) alone. It is a remarkable feature of the other field equations that they conspire, together with (9.42), to ensure the validity of the 'conservation' requirement. I am grateful to Jacob Bekenstein for clarifying this point.

[67] Bekenstein (2004*b*) §III.A.

[68] The T*e*V*e*S theory is the latest in a bunch of bimetric theories of gravitation. Another, due to Nathan Rosen, was designed to avoid the singularities which arise in standard GR; see Rosen (1980). In this theory the metric field $g_{\mu\nu}$ describes gravitation and interacts with matter, so that the line element between neighbouring events $ds^2 = g_{\mu\nu}dx^{\mu}dx^{\nu}$ is measured by way of rods and clocks in the usual way. But a second, background metric field $\gamma_{\mu\nu}$, of constant curvature, exists which serves to define a fundamental rest frame of the universe, and which is capable of entering into the field equations. Its line interval has no direct operational significance. As with T*e*V*e*S, the rods and clocks know which metric to survey because of the way matter fields are postulated to couple to one of them. See also in this connection the 2001 bimetric theory of Drummond, which like Bekenstein's, attempts to eliminate dark matter; Drummond (2001).

This is because its properties may be measured and studied using a variety of different experimental devices, composed of different non-gravitational fields and particles, and, because of universal coupling, the results will be independent of the device. Thus, for instance, the proper time between two events is characteristic of spacetime and of the location of the events, not of the clocks used to measure it.[69]

In the context of the Bekenstein T*e*V*e*S theory, the same reasoning leads inexorably to $\tilde{g}_{\mu\nu}$ being regarded a property of space-time, not $g_{\mu\nu}$. This conclusion is fine as far as it goes, but the message from Chapters 2 and 8, that space-time geometry *per se* does not *explain* the kind of universality in question, should not be forgotten. It is true that in both GR and T*e*V*e*S all matter fields couple in the same way to a single 'metric' field of Lorentzian signature, which ensures local Lorentz covariance; and by coupling minimally, special relativity in its full glory appears to be valid locally. But to say that these metric fields are space-time itself, or properties of space-time, is simply to re-express this remarkable double claim, not to account for it.

And what does special relativity mean in this context? The heart of this theory that is laid bare in GR is, as we saw in the previous section, the combination of the big principle governing the non-gravitational interactions and the fact that curvature terms make no appearance in those equations. It is the dynamics that count. Einstein's 1905 principle theory route to the big principle now more than ever seems like little more than a means, albeit a delicately contrived one, to an end. The thermodynamic template, so important to Einstein in the theoretical conditions that prevailed in 1905 is a ladder that can be kicked away. The dynamical approach to special relativistic kinematics expressed by the unconventional voices, and particularly those of W. Swann, L. Jánossy, and J. S. Bell, recorded in Chapter 7, and further defended in Chapter 8, is in this sense consistent with the spirit of GR and of alternative theories of gravity that preserve the strong equivalence principle. It is because of minimal coupling and local Lorentz covariance that rods and clocks, built out of the matter fields which display that symmetry, behave as if they were reading aspects of the metric field and in so doing confer on this field a geometric meaning. That light rays trace out null geodesics of the field is again a consequence of the strong equivalence principle, which asserts that locally Maxwell's equations of electrodynamics are valid.

But of course Maxwell's equations are not strictly valid locally, and this construal of the minimal coupling principle is only approximately valid. Maxwell's equations represent the classical approximation in quantum electrodynamics, and as we saw in section 9.4.1, very small deviations from null cone behaviour are predicted in QED in special circumstances, such as in a Casimir cavity. This should not entirely surprise us. We only have to remember the role that Einstein's third 1905 paper on Brownian motion, and the experimental confirmation by Perrin in 1908 of Einstein's predictions, played in establishing the limits of validity of the laws

[69] Will (2001), §3.1.1.

of thermodynamics. It is ironic that it was the work of Einstein, amongst others, which established the existence of non-thermodynamic fluctuation phenomena based on constructive theory, given his principle theory approach to SR inspired by thermodynamics. The QED effects mentioned above are the analogues of Brownian motion in space-time theory.

In a sense, SR is not a theory about the behaviour of rods and clocks, because as Einstein stressed their very existence, at least in their familiar forms, is not guaranteed by the big principle. If, however, the Lorentz transformations are to have their usual operational meaning in the context of physical boosts, and if the Minkowski space-time interval is to have some connection with the universal behaviour of rods and clocks—those special 'moving atomic configurations' to repeat yet again Einstein's phrase—ultimately some appeal to quantum theory must be made. In SR, this raises no problem in principle: the big principle after all does not pin down the precise form of the dynamics of matter.[70] But the situation is more complicated in GR. There is, after all, a crucial tension between the dynamical interpretation of SR and the structure of GR. It arises from the fact that Einstein's field equations refer to classical fields. In particular, the matter fields appearing in the definition of the stress-energy tensor are classical. To follow the history of attempts to circumvent this problem would take us too far afield,[71] but the arguments of this book, if correct, can only reinforce the importance of solving it.

[70] A justification of this optimistic view concerning the marriage of SR and quantum mechanics is found in Appendix B.

[71] See the Introduction in Callender and Huggett (2001) for a brief review of responses to this problem.

APPENDIX A

Einstein on General Covariance

Many readers of Einstein's 1916 review paper on GR must have been bewildered by the number of reasons he gave in favour of the principle of general covariance.[1] Einstein cites both Mach's principle and the weak equivalence principle in section 2 of the paper, and the lack of operational significance of coordinate differences for rotating frames, as well as the coordinate-independence of physical happenings in section 3. (It is ironic that the last argument, based on the insight of the reality of 'point-coincidences', turns Einstein's 1915 solution of the underdetermination problem posed by the principle of general covariance into a justification of that principle.) It is clear that Einstein's instinctive feel for the importance of the principle was still outstripping his ability to articulate its fundamental motivation. It appears furthermore that Einstein was still viewing the principle as an 'extension' of the traditional relativity principle shared (as he correctly said) between classical mechanics and SR. This confused idea accompanied his early discussions of the weak equivalence principle and was finally disowned by him in 1926.

It is striking that Einstein did not appeal to the very simple fact that inertial coordinate systems are only defined locally in curved space-time. Of course, to do so is implicitly to acknowledge that inertial coordinates are privileged, but only when gravitational effects can be ignored. Since GR is the theory of gravity (most importantly gravitational tidal effects) it is clear that use must be made of coordinate systems which extend beyond neighbourhoods of events in which the observed geometry is approximately that of the tangent space.[2] But even then, depending on the problem under investigation some coordinate systems are more 'equal' than others.[3] This situation is not novel; it arises in electrodynamics where a number of special gauges are on offer (Lorentz, Coulomb, temporal, etc.) and where choice reflects the problem at hand. The electrodynamic field equations, in their full glory, are standardly written in gauge-general form, but in practice the use of such special gauges abounds.

In his 1916 review paper, Einstein did not think a coordinate-general approach to the field equations was the best route. He promised an 'important simplification of the laws of nature' produced by the restriction to unimodular coordinates, for which $\sqrt{-g} = 1$ (recall the discussion in section 9.2.1 above). Towards the end of the paper, he stated that

[1] Einstein (1916*a*). [2] See Ryckman(2005), note 14, p. 252.

[3] I am struck for example by the usefulness of the harmonic coordinates (which satisfy the de Donder gauge condition $\left(\sqrt{(-g)}g^{\alpha\beta}\right)_{,\beta} = 0$) in the treatment of the Cauchy, or initial value problem in GR (see York (1979)), and in formulating the so-called relaxed form of the field equations (see Will (2001), §4.3 and Wald (1984), p. 261 in this connection). Fock went so far as to say that without appeal to such coordinates the debate between Copernicanism and the Ptolemaic system is vacuous; see Fock (1969), p. 4 and sections 92–3. A rather feeble response to Fock is found in Anderson (1964). For further discussion of the role of special coordinate systems in GR see Zalaletdinov *et al.* (1996).

he had indeed achieved a 'considerable simplification of the formulae and calculations' using these coordinate systems, all in a manner consistent with general covariance![4]

Yet by the time Einstein addressed the famous challenge of Kretschmann over the principle of general covariance in 1918, his thinking had changed. Recall the nature of the challenge: Kretschmann explained that *any* theory could be formulated generally covariantly.[5] The principle has to do with mode of description, not content—it could not be a defining characteristic of Einstein's metric theory of gravity. Einstein had of course to agree, but in addressing Kretschmann's paper, he nonetheless argued that there is good reason why general covariance had proved to have 'considerable heuristic force' in his own work on gravitation.[6] This reason has to do with an interpretation of the principle that transcends Kretschmann's concerns. The principle for Einstein, which he reiterated decades later in his *Autobiographical Notes*, was not just that a theory should have a coordinate general formulation, but that it be such that this formulation is the simplest and most transparent one available to it.

Although there has been debate in the literature as to precisely what Einstein meant here,[7] one reading is particularly plausible. When he proceeds to cite the case of Newtonian mechanics as being ruled out practically if not theoretically by this principle, the problem with the latter theory seems to be the fixed inertial and metric structure therein. A significant simplification of the dynamical description results when it is restricted to global inertial coordinate systems, which are defined in terms of this structure and its symmetries. It seems Einstein was essentially objecting to the existence of absolute geometric objects of this kind. If this was indeed the core of Einstein's response to Kretschmann in 1918, it was essentially an anticipation of the view of Andrej Trautman and James L. Anderson as to the real content of general covariance, which in turn has been defended in many standard texts on GR.[8] Mere coordinate generality has been left far behind.

But there was an earlier shift in Einstein's thinking that is not well known. In November of 1916, Einstein wrote to Hermann Weyl:

> I also came belatedly to the view that the theory [GR] becomes more perspicuous when Hamilton's scheme is applied and when no restrictions are put on the choice of the frame of reference. It is true that the formulas then become somewhat more complicated but more suitable for applications; for it appears that the free choice of the reference system is advantageous in the calculations. The connection between the general covariance requirement and the conservation laws also becomes clearer.[9]

The motivation for at least the last sentence in this passage is clear. In the autumn of 1916, Einstein had anticipated an important application of what is now often called Noether's second theorem.[10]

[4] Einstein (1916*a*), pp. 130 and 156 of the English translation. [5] Kretschmann (1917).
[6] Einstein (1918*b*). [7] See Norton (1993), sections 5.2 and particularly 5.5.
[8] See Misner, Thorne, and Wheeler (1973), section 12.9; Wald (1984) p. 57; Ohanian and Ruffini (1994), section 7.1. [9] Einstein (1998).
[10] Einstein (1916*b*). Brown and Brading (2002) raised the question as to what Einstein had in mind when writing the 1916 letter to Weyl. Despite referring to Einstein's 1916 paper on the variational formulation of GR, these authors did not realize that the answer to the question lay in this paper; the connection was independently clarified in Janssen and Renn (2003), p. 69.

In the same year as the Kretschmann debate, 1918, Emmy Noether, a young mathematician working in David Hilbert's group in Göttingen, published a celebrated paper on the role of symmetry principles in physics.[11] Actually, her paper did not mention the word symmetry; Noether's focus was on transformations which leave the action for some dynamical system invariant, an exercise in the calculus of variations. Normally, however, such transformations happen also to be dynamical symmetries: they take solutions of the Euler–Lagrange equations of motion into solutions. What Noether showed first, in a very systematic fashion, was the existence of a correlation between such dynamical symmetries and strict conservation principles (or more correctly continuity equations), in the case where the symmetries were of the 'global' variety. Examples are rigid spatio-temporal translations, which depend on space-time *independent* parameters. This correlation was not a new result, and nor was it the main object of Noether's paper, but it is the main source of her fame in physics. Noether's real aim in 1918, inspired by Hilbert's work related to Einstein's metric theory of gravity, was to investigate systematically the consequences of variational symmetries of the 'local' variety, which depend on parameters that vary from point to point in space-time. What Noether showed in her 'second theorem' is that each such symmetry gives rise to a 'Noether identity', essentially a condition on the form of the Lagrangian of the dynamical system that holds independently of the field equations themselves.[12] It was shown soon after 1918 that all of Noether's results follow from the weaker condition of quasi-invariance, or invariance of the action up to a surface term, equivalently invariance of the Lagrangian up to a total divergence.[13]

Consider the special case of the action (9.9), the sum of a purely gravitational action and a matter action. Suppose as usual that the gravitational action S_{grav} depends only on $g_{\mu\nu}$, whereas the matter action S_{matter} depends on the generic matter fields ψ_i and $g_{\mu\nu}$. Let us not be specific as to the form of the associated Lagrangian densities, except to demand that each action *separately* is (quasi-) invariant under general coordinate transformations (9.2).[14] The conservation principle (9.19) can be shown to follow from either of the following two procedures. (The fact that more than one exists is typical of the case of coupled fields.)

1. Take the Noether identity associated with the invariance of the matter action, and apply Hamilton's principle to the matter fields, so that their Euler–Lagrange equations hold.[15]

[11] Noether (1918).

[12] This condition expresses an interdependence between the various Euler expressions associated with the action, i.e. the variational derivatives of the total Lagrangian with respect to the dynamical, or dependent variables. Note that it follows from the nature of the Noether identity that a dynamical system with a local symmetry cannot have Euler–Lagrange equations which all take the Cauchy–Kovalevskaya form mentioned in section 9.2.2.; a kind of underdetermination is after all built into the dynamics.

[13] For recent discussions of Noether's theorems see Brading and Brown (2003), and Brown and Holland (2004).

[14] More correctly, we are interested in the vanishing (up to a possible surface term) of the first-order variation in the relevant action under infinitesimal transformations of the form $x^{\mu} \rightarrow x'^{\mu} = x^{\mu} + \epsilon\xi^{\mu}$ where ξ^{μ} is an arbitrary vector field and ϵ is small.

[15] See Carroll (2004), pp. 435–6. In the treatment of Noether's second theorem in Brown and Brading (2002), the argument would put $E^i = 0$ in equation (23), and use the fact that the connection is compatible with the metric: $g^{\mu\nu}{}_{;\alpha} = 0$.

2. Take the Noether identity associated with the invariance of the gravitational part of the action (for the choice of the Hilbert–Einstein action this is the contracted Bianchi identity (9.7)) and apply the gravitational field equations (the analogue of Einstein's equations, or the Euler–Lagrange equations obtained from the total action with respect to variations in $g^{\mu\nu}$).[16] This is a generalization of the way the conservation law is standardly derived in GR.

The paper Einstein wrote in the autumn of 1916 came very close to the derivation based on procedure 2.[17] Although it was far from the first study by Einstein of gravitational dynamics based on an action, or variational, principle, it is the first of his 1915/1916 papers that deals with gravitational dynamics in arbitrary coordinates. Afterwards Einstein would write contentedly to a number of colleagues that he had clarified the connection between the requirement of general covariance (which he still referred to as the 'relativity postulate' at times) and the conservation laws. The only significant difference between his analysis and that given in procedure 2 is that Einstein actually gave the precise form of the gravitational action, and did not realize that this was unnecessary in the argument.[18] How then did Einstein's derivation differ from the now standard one? Einstein did not show explicitly that the Euler expression related to the Hilbert–Einstein action is actually the Einstein tensor density, and hence that the identity associated with invariance is the contracted Bianchi identity. In fact, in 1916 Einstein was unaware of this identity. On the other hand, the fact that the form of the matter action was irrelevant in his derivation was fully appreciated by Einstein; indeed he regarded it as a sign of the superiority of his approach to gravitational dynamics over Hilbert's, which he regarded as too committed to the electromagnetic world picture.

Up to the time Einstein wrote this paper in late 1916, it appears he was convinced that the unimodular coordinates that he had used in his previous papers had true physical significance, in part because of the belief that in such coordinates no gravitational waves propagated without transport of energy. It may even be that this belief extended into 1917.[19] But it is clear that general covariance took on a new meaning for Einstein once he established its relationship with the conservation laws. And yet a word of warning is in order. As Carroll has stressed in this connection[20] (see also the comments on the Lovelock–Grigore theorems in section 9.2.1), the derivation of the conservation law depends not just on (quasi-)invariance of both the gravitational and matter actions, but on the crucial assumption that no fields other than $g_{\mu\nu}$ appear in the gravitational action.

[16] In Brown and Brading (2002), §V, the pair of Noether identities associated with the invariances of the total action and the matter action are taken, one is subtracted from the other, and the gravitational field equations are applied. This is a complicated version of procedure 2.

[17] A very clear account of Einstein's thinking in this paper, and its significance within the development of GR, is found in Janssen and Renn (2005).

[18] Even if Einstein had thought that the choice of quasi-invariant, first-order gravitational action built only out of $g^{\mu\nu}$ is highly constrained, the proof is non-trivial. Recall the discussion of the Lovelock–Grigore theorems in section 9.2.1.

[19] See Janssen and Renn (2005), p. 63.

[20] Carroll (2004), p. 436.

APPENDIX B

Special Relativity and Quantum Theory

B.1 INTRODUCTION

In this appendix, we briefly consider the question as to whether SR and quantum theory are compatible, or if they are, how to understand what appear to be tensions in their marriage.

Part of the story depends on how SR is to be interpreted. An excessively literal geometrical interpretation of SR is likely to run foul of a construction like the de Broglie–Bohm version of relativistic quantum theory.[1] In this theory, the quantum dynamics are Lorentz covariant, but the sub-quantum dynamics are not. This makes the Minkowski structure of space-time relevant only to the quantum dynamics, and the question seems to arise as to 'which space-time' the de Broglie–Bohm ontology is 'in'. The question is bogus: the full set of dynamical equations is fundamental, and the fact that different parts have different symmetry groups does not mean that the fields and/or particles in question are immersed in a background space-time that is somehow schizophrenic or ambiguous as to its geometric structure.

Another part of the story concerns the issue of entanglement of composite systems in quantum theory, and it is that part which is taken up in this appendix.[2] Given that entanglement is a property of certain quantum states, and relativistic versions of quantum theory exist, what is the problem? The immediate response is that entanglement and relativity seem to run up against one another in the spectre of non-locality. We face the question of whether the special sorts of correlation that quantum mechanics allows between spatially separated systems in entangled states, correlations often simply dubbed 'non-local', can be consistent with the strictures of relativity. We should begin, however, by noting that the relations of both entanglement and relativity to non-locality are rather subtle, and that the notion of non-locality itself is rather vague.

The interesting properties of entangled states were emphasized by Einstein, Podolsky, and Rosen (EPR) in their famous 1935 paper;[3] the term 'entanglement' itself was coined by Schrödinger in his work on the quantum correlations that was stimulated by EPR. With the rapid development of quantum information theory over the last 10–15 years and the recognition that entanglement can function as a communication resource, great strides have been made in understanding the properties of entangled states, in particular, with the development of quantitative theories of bipartite (two-party) entanglement and the recognition of qualitatively distinct forms of entanglement. Entanglement assisted communication, an important novel aspect of quantum information theory, also provides

[1] See, e.g., Holland (1993), chap. 12.

[2] This appendix is a shortened version of the recent study of entanglement and special relativity by Timpson and Brown (2002).

[3] Einstein *et al.* (1935).

a new context in which the relations between entanglement, non-locality, and relativity may be explored.

In the following section, the question of how entanglement is related to the notions of non-locality arising from the work of Bell is reviewed, and how this impinges on the constraints of relativity considered. Section B3 is concerned with the twin use of entanglement and wavefunction collapse in Einstein's incompleteness argument, recalling the interesting fact that Einstein's true concern in the EPR argument appears not to have been with relativity. It is often asserted that the Everett interpretation provides an understanding of quantum mechanics unblemished by any taint of non-locality; this claim is also assessed below in section B4.

B.2 ENTANGLEMENT, NON-LOCALITY, AND BELL INEQUALITIES

A state is called entangled if it is not separable, that is, if it cannot be written in the form:

$$|\Psi\rangle_{AB} = |\phi\rangle_A \otimes |\psi\rangle_B, \text{ for pure, or } \rho_{AB} = \textstyle\sum_i \alpha_i \rho_A^i \otimes \rho_B^i, \text{ for mixed states,}$$

where $\alpha_i > 0$, $\sum_i \alpha_i = 1$, and A, B label the two distinct subsystems. The case of pure states of bipartite systems is made particularly simple by the existence of the Schmidt decomposition—such states can always be written in the form:

$$|\Psi\rangle_{AB} = \sum_i \sqrt{p_i}\, |\bar{\phi}_i\rangle_A \otimes |\bar{\psi}_i\rangle_B. \tag{B.1}$$

where $\{|\bar{\phi}_i\rangle\}$, $\{|\bar{\psi}_i\rangle\}$ are orthonormal bases for systems A and B respectively, and p_i are the (non-zero) eigenvalues of the reduced density matrix of A. The number of coefficients in any decomposition of the form is fixed for a given state $|\Psi\rangle_{AB}$. Hence if a state is separable (unentangled), there is only one term in the Schmidt decomposition, and conversely.[4] The measure of degree of entanglement is also particularly simple for bipartite pure states, being given by the von Neumann entropy of the reduced density matrix of A or B (these entropies being equal, from).

To see something of the relation of entanglement to non-locality, we now need to say a little about non-locality. One precise notion of non-locality is due, of course, to Bell. Having noted that the de Broglie–Bohm theory incorporates a mechanism whereby the arrangement of one piece of apparatus may affect the outcomes of distant measurements (due to the interdependence of the space-time trajectories of particles in entangled states), Bell posed the question of whether peculiar properties of this sort might be true of any attempted hidden variable completion of quantum mechanics . He was to show (famously) that this is indeed the case.[5]

Consider two measurements, selected and carried out at spacelike separation on a pair of particles initially prepared in some state and then moved apart. The outcomes of the

[4] For the mixed state case, this simple test does not exist, but progress has been made in providing necessary and sufficient conditions for entanglement for certain systems e.g. $2 \otimes 2$ and $2 \otimes 3$ dimensional systems (see Horodecki *et al.* (2001) for a review).

[5] Bell (1964).

measurements are denoted by A and B, and the settings of the apparatuses by $\mathbf{a}$ and $\mathbf{b}$. (We call this scenario a Bell-type experiment.) We now imagine adding parameters λ to the description of the experiment in such a way that the outcomes of the measurements are fully determined by specification of λ and the settings $\mathbf{a}$ and $\mathbf{b}$. We also make an assumption of locality, that the outcome of a particular measurement depends *only* on λ and on the setting of the apparatus doing the measuring. That is, the outcomes are represented by functions $A(\mathbf{a}, \lambda), B(\mathbf{b}, \lambda)$; the outcome A does not depend on the setting $\mathbf{b}$, nor B on $\mathbf{a}$. The parameter (or 'hidden variable') λ is taken to be chosen from a space Λ with a probability distribution $\rho(\lambda)$ over it, and expectation values for these pairs of measurements will then take the form

$$E(\mathbf{a}, \mathbf{b}) = \int A(\mathbf{a}, \lambda)B(\mathbf{b}, \lambda)\rho(\lambda)d\lambda. \tag{B.2}$$

This sort of theory is known as a deterministic hidden variable theory. From the expression (A.2), a variety of inequalities (Bell inequalities) for the observable correlations in pairs of measurements follow. Violation of such an inequality implies that the correlations under consideration cannot be explained by a deterministic hidden variable model without denying the locality condition and allowing the setting of one apparatus to affect the outcome obtained by the other. It turns out that the quantum predictions for appropriately chosen measurements on, for example, a singlet state, violate a Bell inequality; and the implication is that any attempt to model the quantum correlations by a deterministic hidden variable theory must invoke some non-local mechanism that allows the setting of an apparatus to affect (instantaneously) the outcome of a distant experiment (analogously to the situation in the de Broglie–Bohm theory). Entanglement is a necessary condition for Bell inequality violation, so it is entanglement that makes quantum mechanics inconsistent with a description in terms of a local deterministic hidden variable theory. This is one sense of non-locality.

The discussion of Bell inequalities was later generalized to the case of *stochastic* hidden variable theories, leading to a somewhat different notion of non-locality. In such a stochastic theory, specification of the variables λ only determines the *probabilities* of measurement outcomes.

Bell begins his discussion[6] with an intuitive notion of *local causality*, that events in one spacetime region cannot be causes of events in another, spacelike separated, region. He then goes on to define a model of a correlation experiment as being *locally causal* if the probability distribution for the outcomes of the measurements factorises when conditioned on the 'hidden' state λ in the overlap of the past light cones of the measurement events. Thus the requirement for a locally causal theory is that

$$p_{\mathbf{a},\mathbf{b}}(A \wedge B|\lambda) = p_{\mathbf{a}}(A|\lambda)p_{\mathbf{b}}(B|\lambda),$$

where, as before, A and B denote the outcomes of spacelike separated measurements and $\mathbf{a}, \mathbf{b}$ the apparatus settings.

Imposing this requirement amounts to saying that once all the possible common causes of the two events are taken into account (which, guided by classical relativistic intuitions, we take to reside in their joint past), we expect the probability distributions for the

[6] Bell (1976*b*).

measurement outcomes to be independent and no longer display any correlations.[7] Again, from the assumption of factorisability, a number of inequalities can be derived which are violated for some measurements on entangled quantum states,[8] leading to the conclusion that such correlations cannot be modelled by a locally causal theory. In fact, (note that we could take λ simply to be the quantum state of the joint system) we know that quantum mechanics itself is not a locally causal theory, as it can be seen directly that factorisability will fail for some measurements on entangled states.

However, it is important to note that failure of local causality in Bell's sense does not entail the presence of non-local causes. In arriving at the requirement of factorizability it is necessary to assume something like Reichenbach's principle of the common cause; namely, to assume that if correlations are not due to a direct causal link between two events, then they must be due to common causes, such causes having being identified when conditionalization of the probability distribution results in statistical independence. (Then, arguing contrapositively, if the correlations can't be due to common causes, we can infer that they must be due to direct causes.) Thus although failure of factorizability may imply that quantum correlations cannot be explained by a common cause, it need not imply that there must therefore be direct (and hence non-local) causal links between spacelike separated events: it could be the principle of the common cause that fails. Perhaps it is simply not the case that in quantum mechanics, correlations are always apt for causal explanation.[9] When discussing non-locality in the context of Bell inequality violation, then, we see that it is important to distinguish between the non-locality implied for a deterministic hidden variable model of an experiment and a violation of local causality, which latter need not, on its own, imply any non-locality.

Bell's condition of factorizability is usually analysed as the conjunction of two further conditions, sometimes called *parameter independence* and *outcome independence*. Parameter independence states that the probability distribution for local measurements, conditioned on the hidden variable, should not depend on the setting of a distant measuring apparatus; it is a consequence of the *no-signalling* theorem (which we shall discuss further in the next section) that quantum mechanics, considered as a stochastic hidden variable theory, satisfies this condition. The condition of outcome independence, on the other hand, states that when we have taken into account the settings of the apparatuses on both sides of the experiment and all the relevant factors in the joint past, the probability for an outcome on one side of the experiment should not depend on the actual result at the other. It is this condition that is typically taken to be violated in orthodox quantum mechanics.

[7] The correlation coefficient between two random variables x and y is given by the covariance of x and y, $\mathrm{cov}(x,y) = \langle (x - \langle x \rangle)(y - \langle y \rangle) \rangle$, divided by the square root of the product of the variances. If x and y are statistically independent, $p(x \wedge y) = p(x)p(y)$, then $\mathrm{cov}(x,y) = 0$ and the variables are uncorrelated.

[8] If we wanted to be more precise, we would need also to assume, for example, that the types of measurement A and B chosen did not depend in a conspiratorial way on λ.

[9] In contrast to the present approach, Maudlin insists that failure of factorizability does imply non-local causation as he adheres to the common cause principle, taking it be almost a tautology that lawlike prediction of correlations is indicative of a causal connection, either directly or via a common cause; see Maudlin (2002), ch. 4. One might be worried, however, that this sort of argument does not provide a sufficiently robust notion of cause to imply a genuine notion of non-local action-at-a-distance. Dickson argues that in the absence of full dynamical specification of a model of Bell experiments (of the sort that the de Broglie–Bohm theory, for instance, provides), as opposed to the rather schematic hidden variable schemes we have been considering, it is in any case precipitate to reach much of a conclusion about whether genuine non-locality is involved in a model; see Dickson (1998).

We shall discuss outcome independence further in the context of the Everett interpretation below.[10]

It was pointed out above that entanglement was a necessary condition for Bell inequality violation; somewhat surprisingly, it turns out that the converse is not true. Werner showed that a local hidden variable model could be constructed for a certain class of entangled mixed states ('Werner states'),[11] which means that entanglement is not a sufficient condition for non-locality in the sense of Bell inequality violation; the two terms are not synonymous (although it was subsequently established that entanglement does always imply violation of some Bell inequality for the restricted case of bipartite pure states). The plot was further thickened when Popescu demonstrated that the Werner states for dimensions greater than or equal to five could, however, be made to violate a Bell inequality if sequential measurements are allowed, rather than the single measurements of the standard Bell inequality scenario; this property he termed 'hidden non-locality'.[12] We see that the relationship between entanglement and Bell inequality violation is indeed subtle, just as the links between Bell inequality violation and non-locality are complex. What, then, of relativistic concerns?

Normally one would vaguely assent to the idea that special relativity implies that the speed of light is a limiting speed, in particular, the limiting speed for causal processes or for signalling. But we have seen that the fundamental role of the speed of light in relativity is as the invariant speed of the theory, not, in the first instance, as a limiting speed. That being said, one does still need to pay some attention to the question of superluminal signalling.

The immediate concern with superluminal signals in a relativistic context is familiar; in at least some frames, a superluminal signal is received before it is emitted, so we could arrange for a signal to be sent back that would result in the original signal not being transmitted, hence paradox. Clearly, such signal loops must be impossible. In an influential discussion, though, Maudlin draws a distinction between superluminal signals *simpliciter* and superluminal signals that allow loops, arguing that it is only the latter that need give rise to inconsistencies with relativity.[13] And cases where photons propagate at superluminal speeds in a consistent fashion were touched on in section 9.4.1 above. The fundamental relativistic constraint is that of Lorentz covariance.

How, then, do the notions of non-locality that we have seen to be associated with some forms of entanglement in the violation of Bell inequalities relate to relativity? Only indirectly, at best. Violation of Bell's notion of local causality does not on its own imply non-local causation, hence there is no immediate suggestion of any possible conflict with Lorentz covariance; and although the implied non-local mechanism in a deterministic hidden variable theory would pick out a preferred inertial frame and violate Lorentz covariance[14] this would only be of any significance if we were actually to choose to adopt such a model.[15]

[10] For a discussion of the relation between outcome independence and Reichenbach's principle of the common cause, see Brown (1991).

[11] Werner (1989).

[12] Popescu (1995).

[13] Maudlin (2002).

[14] At least when a full dynamical specification of the theory was forthcoming; cf. the comment on Dickson, fn. 5.

[15] Even then, surprisingly, one might still argue that there is no real violation of relativity; some have suggested that it is only *observable*, or empirically accessible quantities whose dynamics need be Lorentz invariant. This would imply that even the de Broglie–Bohm theory might be consistent with relativity, despite its manifest non-locality.

A more direct link between entanglement and relativity would be in the offing, however, if we were to take collapse of the wavefunction seriously, a matter we have not so far addressed. Collapse seems just the sort of process of action-at-a-distance that would be inimical to relativity, and in the context of EPR or Bell-type experiments, the processes of collapse that would be needed to explain the observed correlations raise profound difficulties with the requirement of Lorentz covariance. Some Lorentz covariant theories of dynamical collapse have been proposed, but rather than consider these, our attention now turns instead to consider how relativity and collapse figure in the context of Einstein's celebrated discussions of the interpretation of quantum theory, before we go on to see how the issue of non-locality is transformed if one denies collapse, as in the Everett theory.

B.3 EINSTEIN, RELATIVITY, AND SEPARABILITY

Einstein's scepticism about quantum mechanics, or at least his opposition to the quantum theory as preached by the Copenhagen school, is well known. As the father of relativity and in the context of the EPR paper, it is tempting to see this opposition as based on his recognizing that the sort of correlations allowed by entangled states will conflict with relativity at some level. At the 1927 Solvay Conference, Einstein did indeed convict quantum theory, considered as a complete theory of individual processes, on the grounds of conflict with relativity. At this early stage, though, his concern was with the action-at-a-distance implied by the collapse on measurement of the wavefunction of a single particle diffracted at a slit (a process that somehow stops the spatially extended wavefunction from producing an effect at two or more places on the detecting screen), rather than anything to do with entanglement.

In fact, as several commentators have remarked, Einstein's fundamental worry in the EPR paper, a worry more faithfully expressed in his later expositions of the argument, was not directly to do with relativity. Rather, his opposition to quantum theory was based on the fact that, if considered complete, the theory violates a principle of separability for spatially separated systems. What he had in mind is expressed in the following passage:

> If one asks what, irrespective of quantum mechanics, is characteristic of the world of ideas of physics, one is first of all struck by the following: the concepts of physics relate to a real outside world, that is, ideas are established relating to things such as bodies, fields, etc., which claim 'real existence' that is independent of the perceiving subject... It is further characteristic of these physical objects that they are thought of as arranged in a space-time continuum. An essential aspect of this arrangement of things in physics is that they lay claim, at a certain time, to an existence independent of one another, provided these objects 'are situated in different parts of space'. Unless one makes this kind of assumption about the independence of the existence (the 'being thus') of objects which are far apart from one another in space, which stems in the first place from everyday thinking, physical thinking in the familiar sense would not be possible. It is also hard to see any way of formulating and testing the laws of physics unless one makes a clear distinction of this kind.
>
> ...The following idea characterizes the relative independence of objects far apart in space (A and B): external influence on A has no direct influence on B; this is known as the 'principle of contiguity', which is used consistently only in the field theory. If this axiom were to be completely abolished, the idea of the existence of (quasi-)enclosed systems, and thereby the postulation of laws which can be checked empirically in the accepted sense, would become impossible.[16]

[16] Einstein (1948).

In this passage, Einstein's description of the grounds for a principle of separability takes on something of the form of a transcendental argument: separability is presented as a condition on the very possibility of framing empirical laws. From the EPR argument, however, it follows that separability is not consistent with the thought that quantum mechanics is complete.

If we consider two entangled systems, A and B, in a pure state, the type of measurement made on A will determine the (pure) state that is ascribed to B, which may be at spacelike separation. Any number of different measurements could be performed on A, each of which would imply a different final state for B. From the principle of separability, however, the real state of a distant system cannot be affected by the type of measurement performed locally or indeed on whether any measurement is performed at all; separability then requires that the real state of B (the physically real in the region of space occupied by B) has *all* of these different possible quantum states associated with it simultaneously, a conclusion clearly inconsistent with the wavefunction being a *complete* description[17].

Since the principle of separability is supported by his quasi-transcendental argument, Einstein would seem to be on firm ground in denying that quantum mechanics can be a complete theory. On further consideration, however, the argument for separability can begin to look a little shaky. Isn't quantum mechanics itself a successful empirical theory? Is it not then simply a counter-example to the suggestion that physical theory would be impossible in the presence of non-separability? As Maudlin puts it: 'quantum mechanics has been precisely formulated and rigorously tested, so if it indeed fails to display the structure Einstein describes, it also immediately refutes his worries.'[18]

As implied in Maudlin's phrasing (note the conditional), some care is required with this response. Since it is precisely the question of how the formalism of quantum mechanics is to be interpreted that is at issue, we are always free to ask *what it is* that is supposed to have been tested by the empirical success of the predictions of quantum theory. A complete and non-separable theory, or an incomplete and statistical one? Appeal to bare empirical success appears insufficient to decide the question of the meaning of the formalism. Let us, then, try to sharpen the intuition that orthodox quantum theory may consitute a counter-example to Einstein's argument.

It is useful to look in a little more detail at the principle of separability. Although it is not clear that he does so in the above quoted passage, elsewhere at least, Einstein distinguished between what might be called separability proper and locality.[19] Separability proper is the requirement that separated objects have their own independent real states (in order that physics can have a subject matter, the world be divided up into pieces about which statements can be made); locality is the requirement that the real state of one system remains unaffected by changes to a distant system. One would usually take

[17] This is the simplified version of the EPR incompleteness argument which Einstein preferred. It is interesting to note the dialectical significance of Einstein's use of entangled systems in the argument for incompleteness. His earlier attempt to argue for incompleteness could be blocked by Bohr's manouevre of invoking the unavoidable disturbance resulting from measurement in order to explain non-classical behaviour and the appearance of probabilities in the quantum description of experiments. Using the correlations implicit in entangled systems to prepare a state cleverly circumvents the disturbance doctrine, as there is no possibility of the mechanical disturbance Bohr had in mind being involved. See Fine (1986), p. 31

[18] Maudlin (1998), p. 49.

[19] See Born *et al.* (1971), p. 164, and Einstein (1969), p. 85.

this locality condition to fail in orthodox quantum mechanics with collapse. The justification for the feeling that the orthodox theory provides a counter-example to Einstein's transcendental argument is that this failure of locality is relatively benign; it does not seem to make the testing of predictions for isolated systems impossible. It is important to note why. The predictions made by quantum mechanics are of the probabilities for the outcomes of measurements. It is established by the no-signalling theorem, however, that the *probabilities* for the outcomes of any measurement on a given subsystem, as opposed to the state of that system, cannot be affected by operations performed on a distant system, even in the presence of entanglement. Thus the no-signalling theorem entails that quantum theory would remain empirically testable, despite violating locality.[20]

There might remain the worry from the Einsteinian point of view that testability at the level of statistical predictions does not help with the problem that the proper descriptive task of a theory may be rendered difficult given a violation of locality. Local experiments may have difficulty in determining the real states, if such there be; but this is a different issue from the question of whether it is possible to state empirical laws. Consider, for example, the de Broglie–Bohm theory. Here we have perfectly coherently stated dynamical laws, yet if the distribution for particle coordinates is given (as usual) by the modulus squared of the wavefunction, it is impossible to know their positions. So it does seem that counter-examples can be given to Einstein's transcendental argument, orthodox quantum theory with collapse being one of them. With this in mind, it is somewhat reassuring that having presented the transcendental argument, Einstein went on to reach a more conciliatory conclusion:

> There seems to me no doubt that those physicists who regard the descriptive methods of quantum mechanics as definitive in principle would... drop the requirement... for the independent existence of the physical reality present in different parts of space; they would be justified in pointing out that quantum theory nowhere makes explicit use of this requirement. I admit this, but would point out: when I consider the physical phenomena known to me... I still cannot find any fact anywhere which would make it appear likely that [the requirement] will have to be abandoned.[21]

Having introduced the no-signalling theorem, we should close this section by remarking that this theorem is of crucial importance in saving quantum mechanics from explicit non-locality; moreover, a form of non-locality that would lead to direct conflict with relativity.[22] For if the probability distribution for measurement outcomes on a distant system could be affected locally, this would provide the basis for superluminal signalling processes that could lead to the possibility of temporal loop paradoxes; and such processes, we have suggested, cannot be incorporated into a Lorentz covariant dynamical theory.

[20] For references to the first versions of the no-signalling theorem, see Timpson and Brown (2002), fn. 12.

[21] Einstein (1948), p. 172.

[22] The no-signalling theorem rules out the possibility of signalling using entangled systems, but note that in the context of non-relativistic quantum mechanics, signalling at arbitrary speeds is nonetheless possible by other means. For example, if the walls are removed from a box in which a particle has been confined, then, instantaneously, there is a non-zero probability of the particle being found in *any* region of space.

B.4 NON-LOCALITY, OR ITS ABSENCE, IN THE EVERETT INTERPRETATION

Einstein's argument for the incompleteness of quantum mechanics made crucial use of the notion of collapse along with that of entanglement. It is interesting to see what happens if one attempts to treat quantum theory *without* collapse as a complete theory. This way lies the Everett approach, to which we shall now turn.

It is a commonplace that the Everett interpretation[23] provides us with a picture in which non-locality plays no part; indeed, this is often presented as one of the main selling points of the approach. Everett himself stated rather dismissively that his relative state interpretation clarified the 'fictitious paradox' of Einstein, Podolsky, and Rosen illustrates in some detail how the absence of collapse in the Everett picture allows one to circumvent the argument for incompleteness.

Two things are important for the apparent avoidance of non-locality in Everett. The first is that we are dealing with a no-collapse theory, a theory of the universal wavefunction in which there is only ever unitary evolution. Removing collapse, with its peculiar character of action-at-a-distance, we immediately do away with one obvious source of non-locality. The second, crucially important, factor (compare the de Broglie–Bohm theory) is that in the Everett approach, the result of a measurement is not the obtaining of one definite value of an observable at the expense of other possible values.

In the presence of entanglement, subsystems of a joint system typically do not possess their own state (i.e. are not in an eigenstate of any observable), but only a reduced density matrix; it is the system as a whole which alone has a definite state. In the Everett approach, however, what claim importance are the states relative to other states in an expansion of the wavefunction. A given subsystem might not, then, be in any definite state on its own, but relative to some arbitrarily chosen state of another subsystem, it *will* be in an eigenstate of an observable. That is, it possesses a definite value of the observable relative to the chosen state of the other system. This allows us to give an explanation of what happens on measurement. Measurement interactions, on the Everett picture, are simply (unitary!) interactions which have been chosen so as to correlate states of the system being measured to states of a measuring apparatus. Following an ideal first-kind (non-disturbing) measurement, the measured system will be in a definite state (eigenstate of the measured observable) relative to the indicator states of the measuring apparatus and ultimately, relative to an observer. If there are a number of different possible outcomes for a given measurement, a state corresponding to each outcome will have become definite relative to *different* apparatus (observing) states, following the interaction. Thus the result of a measurement is not that one definite value alone from the range of possible values of an

[23] It is perhaps worth noting that there have been a number of different attempts to develop Everett's original ideas into a full-blown interpretation of quantum theory (Many Minds, Many Worlds. . .). The most satisfactory of these, in our view, is the approach developed, in slightly different ways, by Simon Saunders and David Wallace, which resolves the preferred basis problem by way of appeal to the phenomenon of decoherence and which has made considerable progress on the question of the meaning of probability in the Everett picture. (See Saunders (1996, 1998) and Wallace (2003*a*, *b*) and further references therein.) It is this version that should be considered, then, when a detailed ontological picture is desired in what follows, although nothing much in the discussion of non-locality should turn on this.

observable obtains, but that each outcome becomes definite relative to a different state of the observer or apparatus.[24]

Given this sort of account of measurement, there appears to be no non-locality, and certainly no conflict with relativity in Everett. The obtaining of definite values of an observable is just the process of one system coming to have a certain state relative to another, a result of local interactions governed by the unitary dynamics. Furthermore, when considering spacelike separated measurements on an entangled system, say measurements of parallel components of spin on systems in a singlet state, there is no question of the obtaining of a determinate value for one subsystem requiring that the distant system acquire the corresponding determinate value, instead of another. Both sets of anti-correlated values are realized (become definite) relative to different observing states; there is, as it were, no dash to ensure agreement between the two sides to be a source of non-locality and potentially give rise to problems with Lorentz covariance.

The distinction introduced above between parameter independence and outcome independence in the framework of stochastic hidden variable theories provides a common terminology in which the oddity of the quantum correlations inherent in entangled states is discussed. As we mentioned, the no-signalling theorem ensures that parameter independence is satisfied for orthodox quantum mechanics, and it is violation of outcome independence that one usually takes to be responsible for failure of factorizability. Recall that the condition of outcome independence states that once one has taken into account the settings of the apparatuses on both sides of the experiment and all the relevant factors in the joint past, the probability for an outcome on one side of the experiment should not depend on the actual result at the other. Violation of this condition, then, presents us with a very odd situation. *For how could it be true of two stochastic processes which are distinct and supposed to be* irreducibly random *that their outcomes may nonetheless display correlations*? (One might say that this question captures the essence of the concerns raised by the EPR argument, but formulated in a way that requires neither explicit appeal to collapse nor to perfect correlations as in the case of parallel spin measurements on the singlet state.)

The Everett interpretation can help us with the oddity of the violation of outcome independence and with understanding how the quantum correlations come about in general. The important thing to note is that the obtaining of the relevant correlations in the Everett picture is not the result of a stochastic process, but of a deterministic one. Given the initial state of all the systems, the appropriate correlations follow deterministically, given the interactions that the systems undergo. Correlations between measurement outcomes obtain if definite post-measurement indicator states of the various measuring apparatuses are definite relative to one another. If we have a complete, deterministic story about how *this* can come about, as we do, then it is difficult to see what more could be demanded in explanation of how the quantum correlations come about. (Of course, the general question

[24] The situation becomes more complicated when we consider the more physically realistic case of measurements which are not of the first kind; in some cases, for example, the object system may even be destroyed in the process of measurement. What is important for a measurement to have taken place is that measuring apparatus and object system were coupled together in such a way that if the object system had been in an eigenstate of the observable being measured prior to measurement, then the subsequent state of the measuring apparatus would allow us to infer what that eigenstate was. In this more general framework the importance is not so much that the object system is left in a eigenstate of the observable relative to the indicator state of the measuring apparatus, but that we have definite indicator states relative to macroscopic observables.

of understanding the nature and capacities of the correlations inherent in entangled states remains, but as a separate issue.)

We have seen how the story goes in the EPR-type scenario: for parallel measurements, given the initial entangled singlet state and given the measurement interactions, it was always going to be the case that an 'up' outcome on side *A* would be correlated with a 'down' outcome on side *B* and a 'down' at *A* with an 'up' at *B* (both outcomes coexisting in the Everett sense). For the case of non-parallel measurements, correlations don't immediately obtain after the two spin measurements, but again, it follows deterministically that the desired correlations obtain if the necessary third measurement is later performed. With regard to the question of outcome independence, we see that we avoid the tricky problem of having to explain how the condition is violated, as from the point of view of Everett, the scenario is not one of distinct stochastic processes mysteriously producing correlated results, but of correlations arising deterministically from an initial entangled state.[25]

[25] For a fuller account, which also contains an analysis of the phenomena of superdense coding and teleportation in the context of the Everett theory, is given in Timpson and Brown (2002).

Bibliography

Ronald Adler, Maurice Bazin, and Menahim Schiffer. *Introduction to General Relativity*. McGraw-Hill Kogakusha Ltd., Tokyo, 2nd edn., (1975). International student edn.

Y. Aharonov and D. Bohm. 'Significance of electromagnetic potentials in quantum theory', *Physical Review*, 115: 485–91 (1959).

Giovanni Amelino-Camilia and Claus Lämmerzahl. 'Quantum-gravity-motivated Lorentz-symmetry tests with laser interferometers', *Classical and Quantum Gravity*, 21: 899–915 (2004).

Jeeva Anandan. 'On the hypotheses underlying physical geometry', *Foundations of Physics*, 10: 601–29 (1980).

—— 'A geometric approach to quantum mechanics', *Foundations of Physics*, 21: 1265–84 (1991).

—— 'The geometric phase', *Nature*, 360: 307–13 (1992).

—— and Harvey R. Brown. 'On the reality of space-time geometry and the wavefunction', *Foundations of Physics*, 25: 349–60 (1995).

James L. Anderson. 'Relativity pirnciples and the role of coordinates in physics', in Hong-Yee Chiu and William F. Hoffmann, eds., *Gravitation and Relativity*, 175–94. W. A. Benjamin Inc., New York (1964).

—— *Principles of Relativity Physics*. Academic Press Inc., New York (1967).

—— 'Newton's first two laws of motion are not definitions', *American Journal of Physics*, 58: 1192–95 (1990).

R. Anderson, I. Vetharanium, and G. E. Stedman. 'Conventionality of synchronization, gauge dependence and test theories of relativity', *Physics Reports*, 295: 93–180 (1998).

Jürgen Audretsch, Friedrich Hehl, and Claus Lämmerzahl. 'Matter wave interferometry and why quantum objects are fundamental for establishing a gravitational theory', in J. Ehlers and G. Shäter, eds., *Relativistic Gravity Research, with emphasis on Experiments and Observations. Lecture Notes in Physics*, 410: 368–407. Springer-Verlag, Berlin, Heidelberg (1992).

John C. Baez and Emory F. Bunn. 'The meaning of Einstein's equation', E-print: arXive:gr-qc/0103044 v3 (2004).

Yuri Balashov and Michel Janssen. 'Critical notice: Presentism and relativity', *British Journal for the Philosophy of Science*, 54: 327–46 (2003).

Julian B. Barbour. 'Relational concepts of space and time', *British Journal for the Philosophy of Science*, 33: 251–74 (1982).

—— *Absolute or Relative Motion? Vol. 1: The Discovery of Dynamics*. Cambridge University Press, Cambridge (1989).

—— *The End of Time: The Next Revolution in Our Understanding of the Universe*. Weidenfeld & Nicolson, London (1999).

—— and Bruno Bertotti. 'Mach's principle and the structure of dynamical theories', *Proceedings of the Royal Society, London*, A 382: 295–306 (1982).

H. Bateman. 'The transformation of the electrodynamical equations', *Proceedings of the London Mathematical Society (2)*, 8: 223–64 (1910).

Jacob D. Bekenstein. 'An alternative to the dark matter paradigm: relativistic MOND gravitation', E-print: arXiv:astro-ph/0412652 v2 (2004*a*).

—— 'Relativistic gravitation theory for the MOND paradigm', E-print: arXiv:astro-ph/0403694 v2 (2004*b*).

John S. Bell. 'On the Einstein–Podolsky–Rosen paradox', *Physics*, 1: 195–200 (1964); repr. in Bell (1987), 14–21.

—— 'How to teach special relativity', *Progress in Scientific Culture*, 1, (1976*a*); repr. in Bell (1987), 67–80.

—— 'The theory of local beables', *Epistemological Letters* (March 1976*b*); repr. in Bell (1987), 52–62.

—— *Speakable and Unspeakable in Quantum Mechanics*. Cambridge University Press, Cambridge (1987).

—— 'George Francis FitzGerald', *Physics World*, 5: 31–35 (1992); 1989 lecture, abridged by Denis Weare.

Gordon Belot. 'Geometry and motion', *British Journal for the Philosophy of Science*, 51: 561–95 (2000).

J. M. Bennett, D. T. McAllister, and G. M. Cabe. 'Albert A. Michelson, Dean of American Optics—life, contributions to science, and influence on modern-day physics', *Applied Optics*, 12: 2253–79 (1973).

Peter G. Bergmann. *Introduction to the Theory of Relativity*. Dover Publications, New York (1976).

—— and Arthur Komar. 'The phase space formulation of general relativity and approaches toward its canonical quantization', in A. Held, (ed.), *General Relativity and Gravitation. One Hundred Years After the Birth of Albert Einstein. Vol. 1*, 227–54. Plenum Press, New York (1980).

V. Berzi and V. Gorini. 'Reciprocity principle and the Lorentz transformations', *Journal of Mathematical Physics*, 10: 1518–24 (1969).

G. Yu. Bogoslovsky. 'A special-relativistic theory of the locally anisotropic space-time. I: The metric and group of motions of the anisotropic space of events', *Il Nuovo Cimento*, 40B(1): 99–133 (1977).

Alfred M. Bork. 'The "FitzGerald" contraction', *Isis*, 57: 199–207 (1966).

Max Born. *Einstein's Theory of Relativity*. Dover publications, New York (1965); revised edn. prepared in collaboration with Günther Leibfried and Walter Biem.

—— H. Born and A. Einstein. *The Born-Einstein Letters*. Macmillan, London (1971).

K. Brading and E. Castellani, (eds.). *Symmetries in Physics: Philosophical Reflections*. Cambridge University Press, Cambridge (2003).

—— and Harvey R. Brown. 'Symmetries and Noether's theorems', in Brading and Castellani (2003), 89–109.

Hale Bradt. *Astronomy Methods. A Physical Approach to Astronomical Observations*. Cambridge University Press, Cambridge (2004).

K. Brecher and J. L. Yun. Note in *Bulletin of the American Astronomical Society*, 20: 987 (1988).

A. Brillet and J. L. Hall. 'Improved laser test of the isotropy of space', *Physical Review Letters*, 42(9): 549–52 (1979).

Harvey R. Brown. 'Discussion: Does the principle of relativity imply Winnie's (1970) equal passage times principle?' *Philosophy of Science*, 57: 313–24 (1990).

——'Nonlocality in quantum mechanics', *Proceedings of the Aristotelian Society, Suppl.* LXV: 141–59 (1991).

——'Correspondence, invariance and heuristics in the emergence of special relativity', in S. French and H. Kamminga, (eds.), *Correspondence, Invariance and Heuristics; Essays in Honour of Heinz Post*, 227–60. Kluwer Academic Press, Dordrecht (1993); repr. in Butterfield *et al.* (1996).

——'On the role of special relativity in general relativity', *International Studies in the Philosophy of Science*, 11: 67–81 (1997).

——'The origins of length contraction: I the FitzGerald–Lorentz deformation hypothesis', *American Journal of Physics*, 69: 1044–54 (2001). E-prints: arXive:gr-qc/0104032; PITT-PHIL-SCI 218.

——'Michelson, FitzGerald, and Lorentz: the origins of special relativity revisited', *Bulletin de la Société des Sciences et des Lettres de Łódź, Vol LIII; Série: Recherches sur les Déformations*, 34: 23–35 (2003). E-print PITT-PHIL-SCI 987.

——and Katherine Brading. 'General covariance from the perspective of Noether's theorems', *Diálogos (Puerto Rico)*, 79 (2002). E-print:PITT-PHIL-SCI 821.

——and Peter Holland. 'Dynamical vs. variational symmetries: understanding Noether's first theorem', *Molecular Physics (11–12 SPEC. ISS.)*, 102: 1133–9 (2004).

——and Adolfo Maia. 'Light-speed constancy versus light-speed invariance in the derivation of relativistic kinematics', *British Journal for the Philosophy of Science*, 44: 381–407 (1993).

——and Oliver Pooley. 'The origins of the spacetime metric: Bell's Lorentzian pedagogy and its significance in general relativity', in Callender and Huggett (2001), 256–72 (2001). E-print: arXive:gr-qc/9908048.

——'Minkowski space-time: a glorious non-entity', E-prints: arXive: physics/0403088; PITT-PHIL-SCI 1661 (2004). A revised version to appear in *The Ontology of Spacetime*, Dennis Dieks (ed.), Elsevier, 2006.

——and Roland Sypel. 'On the meaning of the relativity principle and other symmetries', *International Studies in the Philosophy of Science*, 9: 235–53 (1995).

S. G. Brush. 'Note on the history of the FitzGerald-Lorentz contraction', *Isis*, 58: 230–2 (1984).

H. A. Buchdahl. 'On a Lagrangian for non-minimally coupled gravitational and electromagnetic fields', *Journal of Physics A: Mathematical and General*, 12(7): 1037–43 (1979).

Jed Z. Buchwald. 'The abandonment of Maxwellian electrodynamics: Joseph Larmor's theory of the electron I, II.' *Arch. Internat. Hist. Sci.*, 31: 135–80; 373–435 (1981).

Tim Budden. 'The relativity principle and the isotropy of boosts', in D. Hull, M. Forbes, and K. Okruhlik, (eds.), *PSA 1992 Vol. 1*, 525–41. Philosophy of Science Association, East Lansing, Michigan (1992).

——'Galileo's ship and spacetime symmetry', *British Journal for the Philosophy of Science*, 48: 483–516 (1997*a*).

——'A star in the Minkowksi sky: anisotropic special relativity', *Studies in the History and Philosophy of Modern Physics*, 28(3): 325–61 (1997*b*).

Jeremy Butterfield, Mark Hogarth, and Gordon Belot, (eds.). *Spacetime*. Dartmouth Publishing Company, Aldershot: International research library of philosophy (1996).

Reginald T. Cahill. 'Absolute motion and gravitational effects', *Apeiron*, 11: 53–111 (2004).

Alice Calaprice, (ed.). *The Quotable Einstein.* Princeton University Press, Princeton NJ. (1996).

A. Calinon. *Etude sur les diverses grandeurs.* Gauthier-Villars, Paris (1897).

Craig Callender. 'An answer in search of a question: "proofs" of the tridimensionality of space', *Studies in History and Philosophy of Modern Physics*, 36: 113–36 (2005).

—— and Nick Huggett, (eds.). *Physics Meets Philosophy at the Planck Scale.* Cambridge University Press, Cambridge (2001).

Marco Mamone Capria and F. Pambianco. 'On the Michelson–Morley experiment', *Foundations of Physics*, 24: 885–99 (1994).

C. Carathéodory. 'Untersuchungen über die Grundlagen der Thermodynamik', *Mathematische Annalen*, 67: 355–86 (1909). English trans. by J. Kestin in J. Kestin (ed.), *The Second Law of Thermodynamics*, Dowden, Hutchinson & Ross; Stroudsburg, PA (1976), 229–56.

Sean Carroll. *Spacetime and Geometry. An Introduction to General Relativity.* Addison Wesley, San Francisco (2004).

Alan Chalmers. 'Galilean relativity and Galileo's relativity', in S. French and H. Kamminga, (eds.), *Correspondence, Invariance and Heuristics; Essays in Honour of Heinz Post*, 189–205. Kluwer Academic Press, Dordrecht (1993).

J. M. D. Coey. 'George Francis Fitzgerald, 1851–1901', http://www.tcd.ie/Physics/History/Fitzgerald/FitzGerald.html. The Millenium Trinity Monday Memorial Discourse (2000).

Richard J. Cook. 'Physical time and physical space in general relativity', *American Journal of Physics*, 72(2): 214–19 (2004).

E. Cunningham. 'The principle of relativity in electrodynamics and an extension thereof', *Proceedings of the London Mathematical Society (2)*, 8: 77–98 (1910).

James T. Cushing. 'Electromagnetic mass, relativity and the Kaufmann experiments', *American Journal of Physics*, 49(12): 1133–49 (1981).

O. Darrigol. 'The electron theories of Larmor and Lorentz: a comparative study', *Historical Studies in the Physical and Biological Sciences*, 25: 265–336 (1994).

—— 'Henri Poincaré's criticism of *fin de siècle* electrodynamics', *Studies in History and Philosophy of Modern Physics*, 26: 1–44 (1995).

P. C. W. Davies. 'Quantum mechanics and the equivalence principle', *Classical and Quantum Gravity*, 21: 2761–71 (2004*a*).

—— 'Transit time of a freely-falling quantum particle in a background gravitational field' Eprint: arXive:quant-ph/0407028 (2004*b*).

Talal A. Debs and Michael L. G. Redhead. 'The twin "paradox" and the conventionality of simultaneity', *American Journal of Physics*, 64(4): 384–92 (1996).

R. C. de Miranda Filho, N. P. Andion, and N. C. A. da Costa. 'First order effects in the Michelson–Morley experiment', *Physics Essays*, 15: 1–25 (2002).

Philippe Dennery and André Krzywicki. *Mathematics for Physicists.* John Weatherhill Inc., Tokyo (1967).

Edward A. Desloge and R. J. Philpott. 'Uniformly accelerated reference frames in special relativity', *American Journal of Physics*, 55(3): 252–61 (1987).

Michael Dickson. *Quantum Chance and Non-locality.* Cambridge University Press, Cambridge (1998).

Dennis Dieks. 'The "reality" of the Lorentz contraction', *Zeitschrift für allgemeine Wissenschaftstheorie*, 15: 33–45 (1984).

P. A. M. Dirac, *General Theory of Relativity*. Princeton University Press, Princeton, NJ (1966).

Robert DiSalle. 'On dynamics, indiscernibility, and spacetime ontology', *British Journal for the Philosophy of Science*, 45: 265–87 (1994).

—— 'Spacetime theory as physical geometry', *Erkenntnis*, 42: 317–37 (1995).

Valery P. Dmitriyev. 'The easiest way to the Heaviside ellipsoid', *American Journal of Physics*, 70(7): 717–18 (2002).

I. T. Drummond. 'Bimetric gravity and "dark matter"', *Physical Review D*, 63: 043503–1–13 (2001).

—— and S. Hathrell. 'QED vacuum polarization in a background gravitational field and its effect on the velocity of photons', *Physical Review D*, 22(2): 343–55 (1980).

John Earman. *World Enough and Space-Time: Absolute versus Relational Theories of Space and Time*. MIT Press, Cambridge, Mass. (1989).

—— and Michael Friedman. 'The meaning and status of Newton's law of inertia and the nature of gravitational forces', *Philosophy of Science*, 40: 329–59 (1973).

—— and John Norton. 'What price substantivalism? the hole story', *British Journal for the Philosophy of Science*, 38: 515–25 (1987).

Arthur S. Eddington. *The Nature of the Physical World*. Cambridge University Press, Cambridge (1928).

—— *The Mathematical Theory of Relativity*. Cambridge University Press, Cambridge, 2nd edn. (1965).

—— *Space, Time and Gravitation. An Outline of the General Theory of Relativity*. Cambridge University Press, Cambridge (1966).

W. F. Edwards. 'Special relativity in anisotropic space', *American Journal of Physics*, 31: 482–9 (1963).

J. Ehlers, F. A. E. Pirani, and A. Schild. 'The geometry of free fall and light propagation', in L. O'Raifertaigh, (ed.), *General Relativity. Papers in Honour of J. L. Synge*, 63–84. Clarendon, Oxford (1972).

Jürgen Ehlers. 'Survey of general relativity theory', in W. Israel, (ed.), *Relativity, Astrophysics and Cosmology*, 1–125. D. Reidel Publishing Company, Dordrecht (1973).

A. Einstein. 'Zur Elektrodynamik bewegter körper', *Annalen der Physik*, 17: 891–921 (1905*a*).

—— 'Ist die Trägheit eines Körpers von seinem Energieinhalt abhängig?' *Annalen der Physik*, 18: 639–41 (1905*b*).

—— 'Bemerkung zur Notiz des Herrn P. Ehrenfest: translation deformierbarer Elektronen und der Flächensatz', *Annalen der Physik*, 23: 206–08 (1907*a*). English trans. in Einstein (1989), Doc. 44, 236–7.

—— 'Relativitätsprinzip und die aus demselben gezogenen Folgerungen', *Jarbuch der Radioaktivität und Elektronik*, 4: 411–62 (1907*b*). English trans. in Einstein (1989), Doc. 47, 252–311.

—— 'Le principe de relativité et ses conséquences dans la physique moderne', *Archives des Sciences Physiques et Naturelles*, 29: 5–28 (1910); republished in Einstein (1993).

—— 'Die Grundlage der allgemeinen Relativitätstheorie', *Annalen der Physik*, 49: 769–822 (1916*a*). English translation by W. Perrett and G. B. Jeffrey in Lorentz *et al.* (1923), 108–64.

A. Einstein. 'Hamiltonsches Princip und allgemeine Relativitatstheorie', *Sitzungberichte der Preussischen. Akad. d. Wissesnschaften*, 1111–16 (1916*b*). English trans. by W. Perrett and G. B. Jeffrey in Lorentz *et al.* (1923), 167–73.

—— *Über dei spezielle und die allgemeine Relativitätstheorie, gemeinverstänlich*. Vieweg, Branuschweig (1917). English trans. of the 5th edn. by R. W. Lawson appeared as *Relativity, the Special and the General Theory: A Popular Exposition*, Methuen, London (1920).

—— 'Comment on Weyl 1918', published at end of Weyl (1918); not included in English translation of same (1918*a*).

—— 'Prinzipielles zur allgemeinen Relativitätstheorie', *Annalen der Physik*, 55: 241–4 (1918*b*).

—— 'What is the theory of relativity?', London *Times* (1919); repr. in Einstein (1982), 227–32.

—— 'Inwiefern läßt sich die moderne Gravitationstheorie ohne die Relativät begründen', *Naturwissenschaften*, 8: 1010–11 (1920).

—— 'Geometrie und erfahrung', *Erweite Fassung des Festvortrages gehalten an der presussischen Akademie* Springer: Berlin (1921). Trans. by S. Bargmann as 'Geometry and Experience' in Einstein (1982), 232–46.

—— *The Meaning of Relativity*. Princeton University Press, Princeton (1922).

—— 'Über den Äther', *Sweirzerische naturforschende Gesellschaft, Verhandlungen*, 105: 85–93 (1924). English trans. in Saunders and Brown (1991), 13–20.

—— 'Elementary derivation of the equivalence of mass and energy', *Bulletin of the American Mathematical Society*, 41: 223–30 (1935).

—— 'Quantenmechanik und wirlichkeit', *Dialectica*, 2: 320–4 (1948). English trans. in Born *et al.* (1971), 168–73.

—— *Out of My Later Years*. Littlefields, Adams & Co., New Jersey (1950).

—— 'H. A. Lorentz, his creative genius and his personality', in G. L. de Haas-Lorentz, (ed.), *H. A. Lorentz. Impressions of his Life and Work*, 5–9. North Holland Publishing Co., Amsterdam (1957).

—— 'Autobiographical notes', in P. A. Schilpp, (ed.), *Albert Einstein: Philosopher-Scientist, Vol. 1*, 1–94. Open Court, Illinois (1969).

—— *Ideas and Opinions*. Crown Publishers Inc., New York (1982).

—— *The Collected Papers of Albert Einstein, Vol. 2, The Swiss Years: Writings, 1900–09 (English translation Suppl.)* M. J. Klein, A. J. Kox, J. Renn, and R. Schulman, (eds.). Princeton University Press, Princeton. Translated by A. Beck (1989).

—— *The Collected Papers of Albert Einstein, Vol. 3, The Swiss Years: Writings 1909–1911*, M. J. Klein, A. J. Kox, J. Renn, and R. Schumann (eds.). Princeton University Press, Princeton (1993).

—— 'Letter to Arnold Sommerfeld, January 14 (1908)', Document 73 in *The Collected Papers of Albert Einstein, Vol. 5, The Swiss Years: Correspondence, 1902–14 (English trans. Suppl.)*, M. J. Klein, A. J. Kox, and R. Schulmann (eds.). Princeton University Press, Princeton. Trans. by A. Beck (1995*a*).

—— *The Collected Papers of Albert Einstein, Vol. 4, The Swiss Years: Writings 1912–14*, M. J. Klein, A. J. Kox, J. Renn, and R. Schumann (eds.). Princeton University Press, Princeton (1995*b*).

—— 'Letter to Hermann Weyl (1918)', Document 278 in *The Collected Papers of Albert Einstein, Vol. 8, The Berlin Years: Correspondence, 1914–18 (English trans. Supp.)*, R. Schulmann *et al.* (eds.). Princeton University Press, Princeton, NJ. Trans. by A. M. Hentschel (1998).

—— B. Podolsky, and N. Rosen. 'Can quantum mechanical description of reality be considered complete?', *Physical Review*, 47 (1935).

Arthur Fine. *The Shaky Game. Einstein, Realism and the Quantum Theory.* University of Chicago Press, Chicago (1986).

G. F. FitzGerald. 'The ether and the earth's atmosphere', *Science*, 13: 390 (1889); repr. in part in Bell (1992).

—— 'Aether and matter', *Electrician*, 20 July 1900; repr. in Larmor (1902), 511–15 (1900*a*).

—— 'The relations between ether and matter', *Nature*, 19 July 1900, 265–6; repr. in Larmor (1902), 505–10 (1900*b*).

—— 'Observation, measurement, experiment: short abstract of methods of induction: measurement of time, mass, length, area and volume', in Larmor (1902), 534–47 (1902).

V. Fock. *The Theory of Space, Time and Gravitation.* Pergamon Press, Oxford (1969); trans. from the Russian by N. Kemmer.

J. G. Fox. 'Experimental evidence for the second postulate of special relativity', *American Journal of Physics*, 30: 297–300 (1962).

Michael Friedman. *Foundations of Space-Time Theories: Relativistic Physics and Philosophy of Science.* Princeton University Press, Princeton (1983).

—— 'Geometry as a branch of physics: Background and context for Einstein's "Geometry and Experience"', in D. B. Malament, (ed.), *Reading Natural Philosophy. Essays in the History and Philosophy of Science and Mathematics*, 193–229. Open Court, Chicago (2002).

Galileo Galilei. *Dialogue Concerning Two Chief World Systems.* University of California Press, Berkeley (1960); trans. by S. Drake.

Peter Galison, *Einstein's Clocks, Poincaré's Maps. Empires of Time.* Hodder & Stoughton, London (2004).

Robert Geroch. *General Relativity from A to B.* University of Chicago Press, Chicago and London (1978).

Michel Ghins and Tim Budden. 'The principle of equivalence', *Studies in the History and Philosophy of Modern Physics*, 32(1): 33–51 (2001).

Carlo Giannoni. 'Relativistic mechanics and electrodynamics without one-way velocity assumptions', *Philosophy of Science*, 45: 17–46 (1978).

—— and Oyvind Gron. 'Rigidly connected accelerated clocks', *American Journal of Physics*, 47(5): 431–5 (1979).

Aleksandar Gjurchinovksi. 'Reflection of light from a uniformly moving mirror', *American Journal of Physics*, 72(10): 1316–24 (2004).

R. T. Glazebrook. 'Obituary for H. A. Lorentz', *Nature (London)*, 121: 287 (1928).

H. F. M. Goenner. 'Theories of gravitation with nonmiminal coupling of matter and the gravitational field', *Foundations of Physics*, 14 (1984).

D. R. Grigore. 'The derivation of Einstein equations from invariance principles', *Classical and Quantum Gravity*, 9: 1555–71 (1992).

V. Guillemin and S. Sternberg. *Symplectic Techniques in Physics*. Cambridge University Press, New York (1984).

B. Guinot. 'Atomic time', in Kovalevsky *et al.* (1989), 379–415 (1989*a*).

B. Guinot. 'General principles of the measure of time: astronomical time', in Kovalevsky *et al.* (1989), 351–77 (1989*b*).

J. Wayne Hamilton. 'The uniformly accelerated reference frame', *American Journal of Physics*, 46(1): 83–9 (1978).

James B. Hartle. *Gravity. An Introduction to Einstein's General Relativity*. Addison Wesley, San Francisco (2003).

D. Hasselkamp, E. Mondry, and A. Scharmann. 'Direct observation of the transversal Doppler-shift', *Zeitschrift für Physik A*, 289: 151–5 (1979).

Barbara Haubold, Hans Joachim Haubold, and Lewis Pyensen. 'Michelson's first ether-drift experiment in Berlin and Potsdam', in Stanley Goldberg and Roger H. Steuwer, (eds.), *The Michelson Era in American Science 1870–1930*, 42–54. American Institute of Physics, New York (1988).

Peter Havas. 'Four-dimensional formulations of Newtonian mechanics and their relation to the special and general theory of relativity', *Reviews of Modern Physics*, 36: 938–65 (1964).

—— 'The early history of the "problem of motion" in general relativity', in *Einstein and the History of Genome Relativity, Einstein Studies, Vol. 1*, Don Howard and John Stachel (eds.). Birkhäuser, Boston/Basel/Berlin (1989), pp. 234–76.

—— and Joshua N. Goldberg. 'Lorentz-invariant equations of motion of point masses in the general theory of relativity', *Physical Review*, 128(1): 398–414 (1962).

Oliver Heaviside. 'The electro-magnetic effects of a moving charge', *Electrician*, 22: 147–8 (1888).

Jan Hilgevoord. 'Time in quantum mechanics: a story of confusion', *Studies in History and Philosophy of Modern Physics*, 36: 29–60 (2005).

Dieter Hils and J. L. Hall. 'Improved Kennedy–Thorndike experiment to test special relativity', *Physical Review Letters*, 64(15): 1697–700 (1990).

Banesh Hoffmann. 'Some Einstein anomalies', in G. Holton and Y. Elkana, (eds.), *Albert Einstein: Historical and Cultural Perspectives, The Centennial Symposium in Jerusalem*. Princeton University Press, Princeton (1982).

Peter Holland. *The Quantum Theory of Motion*. Cambridge University Press, Cambridge (1993).

—— and Harvey R. Brown. 'The non-relativistic limits of the Maxwell and Dirac equations: the role of Galilean and gauge invariance', *Studies in the History and Philosophy of Modern Physics*, 34: 161–87 (2003). E-print: PITT-PHIL-SCI 999.

Giora Hon. 'Is the identification of experimental error contextually dependent? The case of Kaufmann's experiment and its varied reception', in Jed Z. Buchwald, (ed.), *Scientific Practice: Theories and Stories of Doing Physics*, 170–223. University of Chicago Press, Chicago (1995).

G. W. Horndeski. 'Conservation of charge and the Einstein–Maxwell equations', *Journal of Mathematical Physics*, 17(11): 1980–7, (1976).

—— and J. Wainwright. 'Energy-momentum tensor of the electromagnetic field', *Physical Review D*, 16: 1691–701 (1977).

M. Horodecki, P. Horodecki, and R. Horodecki. 'Mixed-state entanglement and quantum communication', in G. Alber *et al.* (eds.), *Quantum Information: An Introduction to Basic Theoretical Concepts and Experiments*. Springer-Verlag, New York (2001). E-print: arXive:quant-ph/0109124.

Jong-Ping Hsu and Yuan-Zhong Zhang. *Lorentz and Poincaré Invariance. 100 Years of Relativity. Advanced Series on Theoretical Physical Science, Vol. 8.* World Scientific, Singapore (2001).

B. J. Hunt. 'The origins of the Fitzgerald contraction', *British Journal for the History of Science*, 21: 61–76 (1988).

Herbert E. Ives. 'Graphical exposition of the Michelson–Morley experiment', *Optical Society of America*, 27: 177–80 (1937*a*).

—— 'Light signals on moving bodies as measured by transported rods and clocks', *Optical Society of America*, 27: 263–73 (1937*b*).

—— and G. R. Stilwell. 'An experimental study of the rate of a moving atomic clock', *Optical Society of America*, 28: 215–26 (1938).

Ted Jacobson and David Mattingly. 'Gravity with a dynamical preferred frame', *Physical Review D*, 64: 024028–1–9 (2001).

O. Jahn and V. V. Sreedhar. 'The maximal invariance group of Newton's equations for a free point particle', *American Journal of Physics*, 69(10): 1039–43 (2001).

Mary B. James and David J. Griffith. 'Why the speed of light is reduced in a transparent medium', *American Journal of Physics*, 60(4): 309–13 (1992).

A. L. Janis. 'Simultaneity and conventionality', in R. S. Cohen and L. Laudan, (eds.), *Physics, Philosophy and Psychoanalysis. Essays in honor of Adolf Grunbaum*, 101–10. Reidel, Dordrecht (1983).

L. Jánossy. *Theory of Relativity based on Physical Reality*. Akadémie Kiadó, Budapest (1971).

Michel Janssen. *A Comparison between Lorentz's Ether Theory and Special Relativity in the light of the Experiments of Trouton and Noble*. PhD thesis, University of Pittsburgh, 1995. Posted on the website of the Max Planck Institute for the History of Science.

—— 'COI stories: Explanation and evidence in the history of science', *Perspectives on Physics*, 10: 457–520 (2002*a*).

—— 'Reconsidering a scientific revolution: The case of Einstein *versus* Lorentz', *Physics in Perspective*, 4: 421–46 (2002*b*).

—— 'The Trouton experiment, $E = mc^2$, and a slice of Minkowski space-time', in Abhay Ashtekar *et al.*, (eds.), *Revisiting the Foundations of Relativistic Physics: Festschrift in Honor of John Stachel*, 27–54. Kluwer, Dordrecht (2003).

—— and Jürgen Renn. 'Untying the knot: How Einstein found his way back to field equations discarded in the Zurich notebook', in Jürgen Renn *et al.*, (eds.), *The Genesis of General Relativity: Documents and Interpretation. Vol. 1. General Relativity in the Making: Einstein's Zurich Notebook*. Kluwer, Dordrecht (2005). Forthcoming.

—— and John Stachel. 'The optics and electrodynamics of moving bodies', to appear in Italian in *Storia della Scienza* 7, S. Petruccioli *et al.* (eds.), Istituto della Enciclopedia Italiana (2003).

F. John. *Partial Differential Equations*. Springer-Verlag, New York (1978).

Matti Kaivola, Ove Poulson, Erling Riis, and S. A. Lee. 'Measurement of the relativistic Doppler shift in neon', *Physical Review Letters*, 54(4): 255–8 (1985).

R. J. Kennedy and E. M. Thorndike. 'Experimental establishment of the relativity of time', *Physical Review*, 42: 400–18 (1932).

Hiroshi Kinoshita and Tetsuo Sasao. 'Theoretical aspects of the Earth rotation', in Kovalevsky *et al.* (1989), 173–211 (1989).

C. Kittel. 'Larmor and the prehistory of the Lorentz transformations', *American Journal of Physics*, 42: 726–9 (1974).

Verlyn Klinkenborg. 'The best clock in the world', *Discover*, 50–7 (June 2000).

J. Kovalevsky, I. I. Mueller, and B. Kolaczek. *Reference Frames in Astronomy and Astrophysics*. Kluwer Academic Publishers, Dordrecht (1989).

U. Kraus. 'Brightness and color of rapidly moving objects: The visual appearance of a large sphere revisited', *American Journal of Physics*, 68 (2000).

E. Kretschmann. 'Über den physikalischen Sinn der Relativitätspostulate, *Annalen der Physik*', 53: 575–614 (1917).

K. Lambeck. 'The Earth's variable rotation: some geophysical causes', in Kovalevsky *et al.* (1989), 241–84 (1989).

Claus Lämmerzahl and Mark P. Haugan. 'On the interpretation of Michelson–Morley experiments', *Physics Letters A*, 282: 223–9 (2001). E-print: arXive:gr-qc/0103052.

Cornelius Lanczos. *The Variational Principles of Mechanics*. Dover Publications Inc., New York, 4th edn. (1970).

Peter T. Landsberg. *Thermodynamics and Statistical Mechanics*. Dover, New York (1990).

L. Lange. *Die geschichtliche Entwickelung des Bewegungsbegriffes und ihr voraussichtliches Endergebnis*. Englemann, Leipzig (1886).

Joseph Larmor. 'On a dynamical theory of the electric and luminiferous medium, Part III', *Philosophical Transactions of the Royal Society of London*, 190A: 205 (1897).

——*Aether and Matter*. Cambridge University Press, Cambridge (1900).

——(ed.). *The Scientific Writings of the late George Francis FitzGerald*. Hodges, Figgis & Co., Dublin (1902).

——'On the ascertained absence of effects of motion through the aether, in relation to the constitution of matter, and on the FitzGerald-Lorentz hypothesis', *Philosophical Magazine*, VII: 621–5 (1904).

——*Mathematical and Physical Papers, Vol. I*. Cambridge University Press, Cambridge (1929).

José Latorre, Pedro Pascual, and Rolf Tarrrach. 'Speed of light in non-trivial vacua', *Nuclear Physics B*, 437: 60–82 (1995).

M. Le Bellac and J.-M. Lévy-Leblond. 'Galilean electromagnetism', *Il Nuovo Cimento*, 14 (1973).

A. R. Lee and T. M. Kalotas. 'Lorentz transformations from the first postulate', *American Journal of Physics*, 43: 434–7 (1975).

J.-M. Lévy-Leblond. 'One more derivation of the Lorentz transformation', *American Journal of Physics*, 44: 271–9 (1976).

Stefano Liberati, Sebastiano Sonego, and Matt Visser. 'Faster-than-*c* signals, special relativity, and causality', *Annals of Physics*, 298: 167–85 (2002).

E. Lieb and J. Yngvason. 'The physics and mathematics of the second law of thermodynamics', *Physics Reports*, 310: 1–96 (1999). Erratum *Physics Reports* 314: 669 (1999). E-print: arXive:cond-mat/9708200.

Dierek-Ekkehard Liebscher. *The Geometry of Time*. Wiley-VCH, Wienhelm (2005).

Oliver Lodge. 'G. F. FitzGerald', *Proceedings of the Royal Society*, 75: 152–60 (1905).

A. A. Logunov. *On the articles by Henri Poincaré 'On the dynamics of the electron'*. Publishing Dept. of the Joint Institute for Nuclear Research, Dubna, Dubna (1995); trans. into English by G. Pontecorvo.

H. A. Lorentz. 'De relative beweging van de aarde en den aether', *Koninklijke Akademie van Wetenschappen te Amsterdam, Wis-en Natuurkundige Afdeeling, Versalagen der Zittingen*, 1: 74–9 (1892); repr. in English trans. 'The relative motion of the earth and the ether' in *Collected Papers*, P. Zeeman and A. D. Fokker (eds.), Nijhjoff, The Hague, 219–23 (1937).

——*Versuch einer Thoerie der electrischen und optischen Erscheinungen in bewegten Körpern*. Brill, Leydon (1895). An English trans. of the sections dealing with the MM experiment is found in Lorentz *et al.* (1923) 3–34.

——'Letter to A. Einstein 1915', trans. by A. J. Fox. Part of Archief H. A. Lorentz, Rijksarchief Noord-Holland, Haarlem, the Netherlands. See also Janssen (1995) 197–8.

——*The theory of electrons and its applications to the phenomena of light and radiant heat*. Teubner, Leipzig, 2nd edn., (1916). A course of lectures delivered at Columbia University, New York, in March and April 1906.

——*Lessen over theoretische natuurkunde aan de Rijks-Universiteit te Leiden gegeven, Vol. 6. Het relativiteitsbeginsel voor eenparige translaties (1910–12)*. Brill, Leiden (1922); English trans. *Lectures on Theoretical Physics*, Vol. 3. Macmillan & Co., London. Page references are to the translation.

——'The determination of the potentials in the general theory of relativity, with some remarks about the measurement of lengths and intervals of time and about the theories of Weyl and Eddington', in his *Collected Papers, Vol. 5*, A. D. Fokker and P. Zeeman (eds.); Nijhoff, The Hague, 363–82 (1937). Originally published in *Proceedings of the Royal Academy of Amsterdam* 29: 383 (1923).

—— A. A. Einstein, H. Minkowski, and H. Weyl. *The Principle of Relativity*. Methuen, London (1923).

D. Lovelock. 'The uniqueness of Einstein field equations in a four-dimensional space', *Arch. Rat. Mech. Anal.*, 33: 54–70 (1969).

John Lucas and Peter Hodgson. *Spacetime and Electromagnetism*. Clarendon Press, Oxford (1990).

David Malament. 'Causal theories of time and the conventionality of simultaneity', *Noûs*, 11: 293–300 (1977).

Giulio Maltese and Lucia Orlando. 'The definition of rigidity in the special theory of relativity and the genesis of the general theory of relativity', *Studies in History and Philosophy of Modern Physics*, 26: 263–306 (1995).

Tim Maudlin. 'Substances and spacetimes: what Aristotle would have said to Einstein', *Studies in History and Philosophy of Science*, 21: 531–61 (1990).

——'Part and whole in quantum mechanics', in E. Castellani, (ed.), *Interpreting Bodies*, 46–60. Princeton University Press, Princeton NJ (1998).

—— *Quantum Non-Locality and Relativity*. Blackwell Publishers Ltd., Oxford, 2nd edn. (2002).

H. Melchor. 'Michelson's ether-drift experiments and their correct equations', in Stanley Goldberg and Roger H. Steuwer, (eds.), *The Michelson Era in American Science 1870–1930*, 96–9. American Institute of Physics, New York (1988).

A. A. Michelson. 'The relative motion of earth and ether', *American Journal of Science*, 22: 120–9 (1881).

—— and E. W. Morley. 'On the relative motion of the earth and the luminiferous ether', *American Journal of Science*, 34: 333–45 (1887).

Mordehai Milgrom. 'MOND—a pedagogical review', Eprint: arXive:astroph/0112069 v1 (2001).

Arthur I. Miller. 'On Einstein, light quanta, radiation, and relativity in 1905', *American Journal of Physics*, 44: 912–23 (1976).

—— *Albert Einstein's Special Theory of Relativity: Emergence (1905) and Early Interpretation (1905–09)*. Addison-Wesley, Reading, Mass., (1981).

Robert Mills. *Space, Time and Quanta: An Introduction to Contemporary Physics*. Freeman, New York (1994).

H. Minkowski. 'Raum und zeit', *Physikalische Zeitschrift*, 10: 104–11 (1909). An English trans. by W. Perrett and G. B. Jeffrey is found in Lorentz *et al.* (1923), 75–91 and Hsu and Zhang (2001), 147–61.

Charles W. Misner, Kip S. Thorne, and John Archibald Wheeler. *Gravitation*. Freeman & Co., San Francisco (1973).

G. Morandi, C. Ferrario, G. Lo Vecchio, G. Marmo, and C. Rubano. 'The inverse problem in the calculus of variations and the geometry of the tangent bundle', *Physics Reports*, 188(3 & 4): 147–284 (1990).

Holger Müller, Sven Herrmann, Claus Braxmaier, Stephan Schiller, and Achim Peters. 'Modern Michelson–Morley experiment using cryogenic optical resonators', *Physical Review Letters*, 391: 020401 (2003).

F. Müller-Hoissen and R. Sippel. 'Spherically symmetric solutions of the non-minimally coupled Einstein–Maxwell equations', *Classical and Quantum Gravity*, 5(11): 1473–88 (1988).

Nancy J. Nercessian. ' "Why wasn't Lorentz Einstein?" An examination of the scientific method of H. A. Lorentz', *Centaurus*, 29: 205–42 (1986).

—— ' "Ad hoc" is not a four-letter word: H. A. Lorentz and the Michelson–Morley experiment', in Stanley Goldberg and Roger H. Steuwer (eds.), *The Michelson Era in American Science 1870–1930*, 71–7. American Institute of Physics, New York (1988).

Graham Nerlich. *The Shape of Space*. Cambridge University Press, Cambridge (1976).

C. Neumann. *Ueber die Principien der Galilei-Newton'schen Theorie*. Teubner, Leipzig (1870).

Isaac Newton. 'De Gravitatione', in A. R. Hall and M. B. Hall (eds.), *Unpublished Scientific Papers of Issac Newton*. Cambridge University Press, Cambridge (1962), pp. 123–46.

—— *The Principia. Mathematical Principles of Natural Philosophy (1687)*. University of California Press, Berkeley (1999); translation of Newton. *Philosophia Naturalis Principia Mathematica* by I. Bernard Cohen and Anne Whiteman.

E. Noether. 'Invariante variationsprobleme', *Göttinger Nachrichten, Math.-phys. Kl.*, 235–57 (1918).

John Norton. 'The hole argument', in Arthur Fine and Jarrett Leplin (eds.), *Proceedings of the 1988 Biennial Meeting of the Philosophy of Science Association, Vol. 2*, 56–64 East Lansing, Michigan (1988); reprinted in Butterfield *et al.* (1996), pp. 285–93.

—— 'Coordinates and covariance: Einstein's view of space-time and the modern view', *Foundations of Physics*, 19: 1215–63 (1989).

—— 'Introduction to the Philosophy of Space and Time', in *Introduction to the Philosophy of Science*. Prentice Hall, Englewood cliffs (1992); reprinted in Butterfield *et al.* (1996), pp. 3–56.

—— General covariance and the foundations of general relativity; eight decades of dispute', *Reports on Progress in Physics*, 56: 791–858 (1993).

—— 'Einstein's special theory of relativity and the problems in the electrodynamics of moving bodies that led him to it', to appear in *Cambridge Companion to Einstein*, M. Janssen and C. Lehner (eds.), Cambridge University Press, Cambridge (2005).

R. M. Nugaev. 'Special relativity as a stage in the development of quantum theory', *Historia Scientiarum*, 34: 57–79 (1988).

Hans C. Ohanian. 'The role of dynamics in the synchronization problem', *American Journal of Physics*, 72: 141–8 (2004).

—— and Remo Ruffini. *Gravitation and Spacetime*. W. W. Norton & Co., New York (1994).

L. B. Okun. 'The concept of mass (mass, energy, relativity)', *Soviet Physics Uspekhi*, 32(7): 629–38 (1989); trans. by Julian B. Barbour, from *Usp. Fiz. Nauk.* 158: 511–30 (July 1989).

Abraham Pais. *'Subtle is the Lord ...' The Science and the Life of Albert Einstein*. Oxford University Press, New York (1982).

—— *The Genius of Science: A Portrait Gallery*. Oxford University Press, Oxford (2000).

Michel Paty. 'Physical geometry and special relativity. Einstein and Poincaré', in J.-M. Salaskis, L. Boi, and D. Flament, (eds.), *1830–1930: A Century of Geometry. Lecture Notes in Physics: 402*. Springer-Verlag, Berlin Heidelberg (1992).

—— 'Poincaré et le principe de relativité', in Kuno Lorenz, Jean-Louis Greffe, and Gerhard Heinzmann, (eds.), *Henri Poincaré. Science et philosophie*, pp. 101–43. Albert Blanchard, Paris (1994).

Wolfgang Pauli. *Theory of Relativity*. Dover, New York (1981). Originally published as 'Relativitätstheorie', *Encyklopädie der matematischen Wissenschaften, mit Einschluss ihrer Anwendungen vol 5. Physik* ed. A. Sommerfeld (Tauber, Leibzig), 539–775 (1921).

K. M. Pedersen. 'Water-filled telescopes and the pre-history of Fresnel's ether dragging', *Archive for the History of the Exact Sciences*, 54: 499–564 (2000).

Herbert Pfister. 'Newton's first law revisited', *Foundations of Physics Letters*, 17 (2004).

Henri Poincaré. 'La mesure du temps', *Revue Métaphysique et de Morale*, 6: 1–13 (1898); English trans. in chap. II of Poincaré 26–36 (1985).

—— 'La théorie de Lorentz et le principe de réaction', *Archives néerlandaises des Sciences exactes et naturelles, 2 ème série*, 5: 252–78 (1900).

—— 'Sur la dynamique de l'électron', *Comptes Rendus de l'Académie des Sciences (Paris)*, 140: 1504–8; (1905); an English trans. is found in Logunov (1995).

—— 'Sur la dynamique de l'électron', *Rendiconti del Circolo matematico di Palermo*, 21: 129–75; 1906; an English trans. is found in Logunov (1995).

—— *Science and Hypothesis*. Dover, New York (1952); trans. by W. Scott of *La Science et l'hypotèse*, E. Flammarion, Paris (1902).

—— *The Value of Science*. Dover, New York (1958); trans. by G. B. Halsted.

Oliver Pooley. 'Handedness, parity violation, and the reality of space', in Brading and Castellani (2003), 250–80. E-print; PITT-PHIL-SCI 221 (2003).

—— 'Points, particles and structural realism', S. French, D. Rickles, and J. Saatsi (eds.), *Structural Foundations of Quantum Gravity*. Oxford University Press, Oxford; in press (2006).

—— and Harvey R. Brown. 'Relationalism rehabilitated? I: Classical mechanics', *British Journal for the Philosophy of Science*, 53: 183–204 (2002). E-print: PITT-PHIL-SCI 220.

S. Popescu. 'Bell's inequalities and density matrices. Revealing "hidden" non-locality', *Physical Review Letters*, 74: 2619–22 (1995). E-print: arXive:quant-ph/9502005.

Karl Popper. *Unended Quest. An Intellectual Autobiography*. Fontana/Collins, Glasgow (1982).

Maxim Pospelov and Michael Romalis. 'Lorentz invariance on trial', *Physics Today*, July: 40–6 (2004).

Michael L. G. Redhead. 'Relativity, causality and the Einstein–Rosen-Podolsky paradox: nonlocality and peaceful existence', in R. Swinburne, (ed.), *Space, Time and Causality*, 151–89. Reidel, Dordrecht (1983).

Dragan V. Redžić. 'Image of a moving spheroidal conductor', *American Journal of Physics*, 60(6): 506–8 (1992).

—— 'Image of a moving sphere and the FitGerald–Lorentz contraction', *European Journal of Physics*, 25: 123–6 (2004).

J. S. Rigden. 'Editorial: High thoughts about Newton's first law', *American Journal of Physics*, 55 (1987).

W. Rindler. 'Einstein's priority in recognizing time dilation physically', *American Journal of Physics*, 38: 1111–15 (1970).

A. A. Robb. *A Theory of Space and Time*. Cambridge University Press, Cambridge (1914).

H. P. Robertson. 'Postulate *versus* observation in the special theory of relativity', *Reviews of Modern Physics*, 21: 378–82 (1949).

John Roche. 'What is mass?', *European Journal of Physics*, 26: 225–42 (2005).

F. Rohrlich. 'The logic of reduction: The case of gravitation', *Foundations of Physics*, 19: 1151–70 (1989).

Nathan Rosen. 'General relativity with a background metric', *Foundations of Physics*, 10(9/10) (1980).

Carlo Rovelli. 'Halfway through the woods: contemporary research on space and time', in John Earman and John D. Norton, (eds.), *The Cosmos of Science*, 180–223. University of Pittsburgh Press, Pittsburgh (1997).

Thomas Ryckman. *The Reign of Relativity. Philosophy in Physics 1915–1925*. Oxford University Press, New York (2005).

Robert A. Rynasiewicz. 'Lorentz' local time and the theorem of corresponding states', in *PSA 1988, vol. 1*, 67–74. Philosophy of Science Association, East Lansing, Michigan (1988).

—— 'The construction of the special theory: some queries and considerations', in Don Howard and John Stachel, (eds.), *Einstein: The Formative Years, 1879–1909. (Einstein Studies, vol. 8)*. Birkhäuser, Boston (2000*a*).

—— 'Definition, convention, and simultaneity: Malament's result and its alleged refutation by Sarkar and Stachel', *Philosophy of Science* 68 (Proceedings), S345–S357. E-print: PITT-PHIL-SCI 125 (2000*b*).

—— 'Is simultaneity conventional despite Malament's result?' Paper given at Humanities and Social Sciences Federation of Canada Congress, May 24–26 Quebec. E-print: PITT-PHIL-SCI 293 (2001).

—— 'The optics and electrodynamics of "On the electrodynamics of moving bodies"', forthcoming in *Annalen der Physik* (2005).

D. H. Sadler. 'Astronomical measures of time', *Quarterly Journal of the Royal Astronomical Society*, 9: 281–93 (1968).

Sahotra Sarkar and John Stachel. 'Did Malament prove the non-conventionality of simultaneity in the special theory of relativity?' *Philosophy of Science*, 66: 208–20 (1999).

Tilman Sauer. 'Albert Einstein's 1916 review article on general relativity', in Ivor Grattan-Guinness, (ed.), *Landmark Writings in Western Mathematics, 1640–1940*. Elsevier, Amsterdam (2005).

Simon W. Saunders. 'Relativism', in Rob Clifton, (ed.), *Perspectives on Quantum Reality*, 125–42. Kluwer Academic Publishers, Dordrecht (1996).

—— 'Time, quantum mechanics, and probability', *Synthese*, 114: 373–404 (1998).

—— and Harvey R. Brown, (eds.). *Philosophy of Vacuum*. Oxford University Press, Oxford (1991).

Bernd Schmidt. 'Mathematics of general relativity', in G. S. Hall and J. R. Pulham, (eds.), *General Relativity*, 1–17. SUSSP Publications and Institute of Physics Publishing, Bristol, (1996). Proceedings of the Forty Sixth Scottish Universities Summer School in Physics, Aberdeen, July 1995.

Erwin Schrödinger. *Space-Time Structure*. Cambridge University Press, Cambridge (1985).

R. A. Schumacher. 'Special relativity and the Michelson–Morley interferometer', *American Journal of Physics*, 62(7): 609 (1994).

W. H. E. Schwartz, A. Rutowksi, and G. Collignon. 'Nonsingular relativistic perturbation theory and relativistic changes of molecular structure', in I. P. Grant, S. Wilson, and B. L. Gyorffy, (eds.), *The Effects of Relativity in Atoms, Molecules, and the Solid State*, 135–47. Plenum Press, New York (1991).

G. F. C. Searle. 'Problems in electrical convection', *Philosophical Transactions of the Royal Society of London, Series A*, 187: 675–713 (1896).

Graham M. Shore. 'Quantum gravitational optics', *Contemporary Physics*, 44: 503–21 (2003).

L. Silberstein. *The Theory of Relativity*. Macmillan, London (1914).

T. Sjödin. 'Synchronization in special relativity and related theories', *Nuovo Cimento B Serie*, 51: 229–46 (June 1979).

Lawrence Sklar. 'Facts, conventions, and assumptions in the theory of space-time', in J. Earman, C. Glymour, and J. Stachel, (eds.), *Foundations of Space-Time Theories. Minnesota Studies in the Philosophy of Science, vol. 8*, 206–74. University of Minnesota Press, Minneapolis (1977).

—— *Philosophy and Spacetime Physics*. University of California Press, Berkeley (1985).

C. Skordis, D. F. Mota, P. G. Ferreira, and C. Boehm. 'Large scale structure in Bekenstein's theory of relativistic MOND', ArXive: astro-ph/0505-519.

Dava Sobel. *Longitude: the true story of a lone genius who solved the greatest scientific problem of his time*. Fourth Estate, London (1996).

Sebastiano Sonego. 'Is there a spacetime geometry?', *Physics Letters A*, 208: 1–7 (1995).

Sebastiano Sonego and Valerio Faraoni. 'Coupling to the curvature for a scalar field from the equivalence principle', *Classical and Quantum Gravity*, 10: 1185–87 (1993).

—— and Hans Westman. 'Particle detectors, geodesic motion, and the equivalence principle', E-print: arXive:gr-qc/0307040 v2 (2003).

V. S. Soni. 'Michelson–Morley analysis', *American Journal of Physics*, 57: 1149–50 (1989).

John Stachel. 'Einstein and Michelson, the context of discovery and the context of justification', *Astronomische Nachrichten*, 303: 47–53 (1982); reprinted in Stachel (2002*a*), pp. 177–90.

—— 'Einstein and the quantum: fifty years of struggle', in Robert G. Colodny, (ed.), *From Quarks to Quasars: Philosophical Problems of Modern Physics*, 349–81. University of Pittsburgh Press, Pittsburgh (1986); reprinted in Stachel (2002*a*), pp. 367–402.

—— 'Einstein's search for general covariance, 1912–1915', in Don Howard and John Stachel, (eds.), *Einstein and the History of General Relativity*. Birkhäuser, Boston (1989*b*); reprinted in Stachel (2002*a*), pp. 301–37.

—— 'History of relativity', in Laurie M. Brown, Abraham Pais, and Brian Pippard, (eds.), *Twentieth Century Physics, vol. 1*, 249–356. American Institute of Physics, New York (1995).

—— (ed.). *Einstein's Miraculous Year: Five Papers that Changed the Face of Physics*. Princeton University Press, Princeton NJ (1998).

—— *Einstein from 'B' to 'Z'*. Birkäuser, Boston/Basel/Berlin (2002*a*).

—— ' "What song the Syrens sang": how did Einstein discover special relativity?', in Stachel (2002*a*), pp. 157–69 (2002*b*).

—— 'Einstein's 1912 manuscript as a clue to the development of special relativity', manuscript (2005).

—— and D. C. Cassidy, J. Renn, and R. Schulmann (eds.), *The Collected Papers of Albert Einstein, vol 2. The Swiss Years: Writings, 1900–1919*. Princeton University Press, Princeton, NJ (1989).

Richard Staley. 'On the histories of relativity: the propagation and elaboration of relativity theory in participant histories in Germany, 1905–1911', *Isis*, 89: 263–99 (1998).

—— 'Travelling light', in Marie-Noëlle Bourguet, Christian Licoppe, and H. Otto Sibum, (eds.), *Instruments, Travel and Science. Itineraries of precision from the seventeenth to the twentieth century*, 243–72. Routledge, London (2002).

—— 'The interferometer and the spectroscope: Michelson's standards and the spectroscopic community', *Nuncius*, 18: 37–59 (2003).

S. Sternberg. 'On the influence of field theories on our physical conception of geometry', in *Proceedings of the Bonn Conference on Differential Geometric Methods in Physics (1977). Springer Lecture Notes in Mathematics, No. 675*, Springer-Verlag, Berlin, New York (1978).

Susan G. Sterrett. 'Sounds like Light: Einstein's special theory of relativity and Mach's work in acoustics and aerodynamics', *Studies in History and Philosophy of Modern Physics*, 29(1): 1–35 (1998).

W. F. G. Swann. 'Relativity and electrodynamics', *Reviews of Modern Physics*, 2(3): 243–304 (1930).

—— 'The relation of theory to experiment in physics', *Reviews of Modern Physics*, 13: 190–6 (1941*a*).

—— 'Relativity, the FitzGerald–Lorentz contraction, and quantum theory', *Reviews of Modern Physics*, 13: 197–202 (1941*b*).

Bryan Sykes. *Adam's Curse*. Corgi Books, Reading (2004).

Roland Sypel and Harvey R. Brown. 'When is a physical theory relativistic?', in *PSA 1992, vol. 1 (Proceedings of the 1992 Biennial Meeting of the Philosophy of Science Association)*, 507–14. Philosophy of Science Association, East Lansing, Michigan (1992).

Laurence G. Taff. *Celestial Mechanics. A Computational Guide for the Practitioner*. John Wiley & Sons, New York (1985).

Reza Tavakol and Roustam Zalaletdinov. 'On the domain of applicability of general relativity', *Foundations of Physics*, 28(2): 307–31 (1998).

Y. P. Terletskii. *Paradoxes in the Theory of Relativity*. Plenum, New York (1968).

P. Teyssandier. 'Variation of the speed of light due to non-minimal coupling between electromagnetism and gravity', E-print: arXive:gr-qc/0303081 v3 (2003).

Christopher Timpson and Harvey R. Brown. 'Entanglement and relativity', in *Understanding Physical Knowledge*, 147–66. Preprint no. 24, Departimento di Filosofia, Università di Bologna, CLUEB, Bologna, (2002). E-print: arXive:quant-ph/0212140.

Roberto Torretti. *Relativity and Geometry*. Pergamon Press, Oxford (1983).

A. Trautman. 'The general theory of relativity', *Soviet Physics Uspekhi*, 89: 319–36 (1966).

Jos Uffink. 'Bluff your way in the second law of thermodynamics', *Studies in History and Philosophy of Modern Physics*, 32B: 305–94 (2001).

William G. Unruh and Robert M. Wald. 'Time and the interpretation of canonical quantum gravity', *Physical Review D*, 40(8): 2598–614 (1989).

Antony Valentini. 'Hidden variables and the large scale structure of spacetime', Eprint: arXive:quant-ph/0504011; to appear in *Absolute Simultaneity*, W. L. Craig and Q. Smith (eds.), Routledge, London (2005).

V. P. Vizgin. *Unified Field Theories in the First Third of the 20th Century*. Birkhäuser Verlag, Basel; trans. by Julian Barbour (1985).

W. Voigt. 'Über das Doppler'sche princip', *Nachrichten der K. Gesellschaft der Wissenschaften zu Göttingen*, 41: 41–51 (1994). An English translation by A. Ernst and J.-P. Hsu is found in Hsu and Zhang (2001), 10–19.

Robert M. Wald. *General Relativity*. Chicago University Press, Chicago (1984).

David Wallace. 'Everett and structure', *Studies in History and Philosophy of Modern Physics*, 34: 87–105 (2003*a*).

—— 'Everettian rationality: defending Deutsch's approach to probability in the Everett interpretation', *Studies in History and Philosophy of Modern Physics*, 34: 415–38 (2003*b*).

Andrew Warwick. 'On the role of the Fitzgerald–Lorentz contraction hypothesis in the development of Joseph Larmor's electronic theory of matter. *Archive for the History of Exact Science*, 43: 29–91 (1991).

R. F. Werner. 'Quantum states with Einstein–Podolsky–Rosen correlations admitting a hidden-variable model', *Physical Review A*, 408: 4277–81 (1989).

S. A. Werner. 'Gravitational, rotational and topological quantum phase shifts in neutron interferometry', *Classical and Quantum Gravity*, 11: A207–A226 (1994).

H. Weyl. 'Gravitation und Elekrizität', *Sitzungsberichte der Königlich Preussische Akademie der Wissenschaften (Berlin) Sitzungsberichte. Physikalisch-Mathematische Klasse*, 465–80; English trans. by W. Perret and G. B. Jeffrey in Lorentz *et al.* (1923), 201–216 (1918).

H. Weyl. *Space-Time-Matter*. Dover, New York (1952); trans. by H. L. Brose of *Raum-Zeit-Materie. Vorlesungen über allgemeine Relativitätstheorie*, Springer, Berlin (1921). Page references to English translation.

Eugene P. Wigner. 'Symmetry and conservation laws', *Physics Today*, 17(3): 34–40 (1964).

Clifford M. Will. 'The confrontation of general relativity and experiment', *Living Reviews of Relativity* 4. [Online article]: cited on January 2005; http://www.livingreviews.org/Articles/Volume4/2001-4will/ (2001).

R. B. Williamson. 'Logical economy in Einstein's "*On the electrodynamics of moving bodies*"', *Studies in History and Philosophy of Science*, 8: 49–60 (1977).

John Winnie. 'Special relativity without one way velocity assumptions I; II', *Philosophy of Science*, 37: 81–99, 223–38 (1970).

C. N. Yang. 'Gauge fields, electromagnetism and the Bohm–Aharonov effect', in S. Kamefuchi, H. Ezaira, Y. Murayama, N. Namiki, S. Nomura, Y. Ohnuki, and T. Yajima (eds.), *Foundations of Quantum Mechanics in the Light of New Technology*. Physical Society of Japan, Tokyo, 5–9 (1984).

—— 'Hermann Weyl's contribution to physics', in K. Chandrasekharan (ed.), *Hermann Weyl 1885–1985*. Springer-Verlag, Berlin, 5–9 (1986).

James W. York. 'Kinematics and dynamics of general relativity', in L. Smarr, (ed.), *Sources of Gravitational Radiation*. Cambridge University Press, Cambridge (1979).

E. G. Zahar. 'Mach, Einstein and the rise of modern science', *British Journal for the Philosophy of Science*, 28: 195–213 (1977).

—— 'Poincaré's independent discovery of the relativity principle', *Fundamenta Scientiae*, 4 (1983).

—— *Einstein's Revolution. A Study in Heuristic*. Open Court, La Salle (1989).

Roustam Zalaletdinov, Reza Tavakol, and George F. R. Ellis. 'On general and restricted covariance in general relativity', *General Relativity and Gravitation*, 28: 1251–67 (1996).

Index